ELEMENTS OF THERMAL TECHNOLOGY

ENGINEERING TECHNOLOGY

A Series of Textbooks

EDITOR

Gerald A. Rath

Wichita State University
Wichita, Kansas

1. Thermodynamics for Engineering Technologists, *Robert K. Will*
2. Engineering Technology Problem Solving: Techniques Using Electronic Calculators, *Houston N. Irvine*
3. Elements of Thermal Technology, *John H. Seely*

Other Volumes in Preparation

ELEMENTS OF THERMAL TECHNOLOGY

JOHN H. SEELY
California State Polytechnic University
Pomona, California

MARCEL DEKKER, INC. New York and Basel

Library of Congress Cataloging in Publication Data

Seely, John H.
Elements of thermal technology.

Includes index.
1. Thermodynamics. 2. Heat -- Transmission.
I. Title.
TJ265.S457 621.402'1 81-1577
ISBN 0-8247-1174-2 AACR2

MARCEL DEKKER, INC.
270 Madison Avenue, New York, New York 10016

Current printing (last digit):
10 9 8 7 6 5 4 3 2 1

PRINTED IN THE UNITED STATES OF AMERICA

PREFACE

All types of engineering systems--mechanical, electrical, thermal, chemical, and fluid--are affected by the manner in which heat is transferred. For example, temperature levels inside electrical components determine the efficiency and effectiveness of circuits made from the components. Many fluid properties such as viscosity, density, and surface tension are highly thermophysical in nature. Strength considerations, friction forces, and clearances between parts are functions of temperature in mechanical systems, while the manner and rate of reaction between the elements of chemical systems also depend on temperature.

The purpose of this book is to outline the elements required by a technologist to analyze or synthesize engineering systems from the viewpoint of thermal technology. Material in this book covers the thermal equilibrium conditions normally treated in classical thermodynamics, and nonequilibrium conditions that constitute the subject of heat transfer. Other sections are devoted to discussions of thermophysical properties, that is, the properties of a system that depend upon temperature.

Although some knowledge of calculus is required to understand the equations used in thermal technology, an effort has been made to minimize the level of mathematics used in presenting the subject matter. Most students who have completed at least one full year of college mathematics should have no difficulty in following the text. Extensive use is made of techniques and empirical equations based on curves, charts, and diagrams.

SI units are used throughout this book. Since many of the references that the technologist will have to use will not be in SI units, appropriate conversion methods are presented where necessary.

The text has been written expressly for students in four-year engineering technology programs. However, it is applicable to the needs of many junior college students and will serve as convenient review material for practicing engineers and engineering technologists.

I am greatly indebted to my wife, Marcella, for excellent typing and proofreading, but most of all for her constant encouragement.

John H. Seely

CONTENTS

1

THERMODYNAMIC DEFINITIONS AND CONCEPTS

1.1 INTRODUCTION

Engineering thermodynamics is the study of energy transfer between systems and the effects of temperature level on the transfer process. Although the subject can be developed from purely macroscopic postulates, knowledge of microscopic concepts helps to develop an understanding of the nature of the matter which is responsible for the macroscopic properties of a substance.

The microscopic approach assumes that matter is particulate and is made up of electrons, protons, neutrons, etc. Each of the particles can be assumed to have a characteristic mass, energy, momentum, and electric charge. The macroscopic approach assumes that matter is continuous and distributed throughout space. A continuum has the same kinds of characteristics as particles, but additional characteristics such as pressure, temperature, mass flow rate, and electric current can also be considered.

1.2 THERMAL EQUILIBRIUM

From the continuum viewpoint, properties such as temperature and pressure are considered to be constant throughout a volume that consists of many different particles; such a consideration implies that the entire system is in equilibrium.

Equilibrium can generally be defined as a condition in which all tendencies for change are equally and oppositely balanced by other tendencies for change. Engineering applications, however, require a more precise definition. An engineering system is said to be in *thermodynamic equilibrium* if it is in mechanical, thermal, electrical, and chemical equilibrium

1. Mechanical equilibrium connotes the absence of any pressure gradients within a system.
2. Thermal equilibrium connotes the absence of any temperature gradients within a system.
3. Electrical equilibrium connotes the absence of any voltage gradients within a system.
4. Chemical equilibrium connotes the absence of any tendency for chemical combination within a system.

The concept of equilibrium presents something of a dilemma. Macroscopic properties, for example, describe the condition of a system when the system is in equilibrium. How, then, can we apply these properties to a system undergoing an energy transfer process, which by definition is a nonequilibrium condition? The problem is resolved by means of an idealization called a *quasistatic* or *quasi-equilibrium* process. Such a process is defined as one in which there is an infinitesimal deviation from thermodynamic equilibrium at each point in the process; however, all states in a quasistatic process can be considered to be equilibrium states.

Consider the system shown in Fig. 1.1 in which the internal pressure of the gas is balanced by the weights; consequently, the piston is stationary. If any of the weights are removed, mechanical equilibrium no longer exists and the piston will rise until equilibrium is reestablished. A quasistatic process can be assumed for the system if the weights are taken to be very small and are removed one at a time. Obviously, if all the weights are removed at once, the piston will rise rapidly and equilibrium will not exist during any part of the process.

Many engineering systems are not too far removed from a quasi-static process, and application of this concept can greatly simplify thermodynamic analysis.

1.3 SYSTEM CONCEPT

A *system* is an entity whose overall performance can be monitored and analyzed. It can be a region in space, a quantity of matter, a group of creatures, or a series of interacting elements. Everything external to the system is the *environment*, and the interface between the system and its environment is the *boundary*. Macroscopic and/or microscopic characteristics of a mechanical, thermal, electrical, or chemical nature can be associated with the system, its boundary, and the environment. These characteristics can change with time or they may be constant while the system is functioning. Outside the realm of technology, the characteristics can be psychological economic, political, or philosophic in nature.

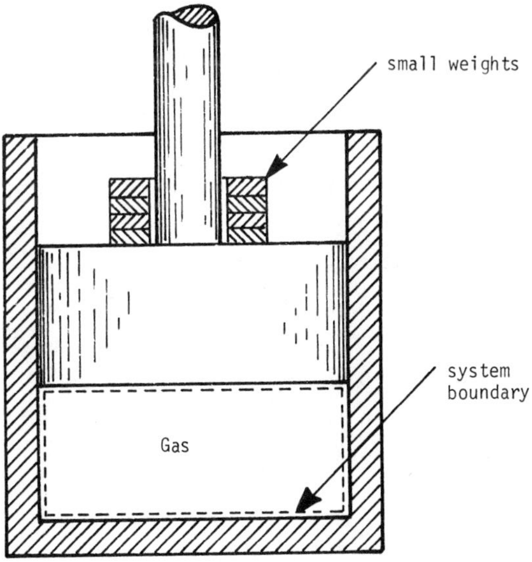

FIGURE 1.1 Gas in the cylinder undergoes a quasistatic process when small weights are removed one at a time.

A *subsystem* is an assemblage of interrelated components, or other subsystems, which make up a system. A *component* is an element which is used to build systems or subsystems, but which does not have to be designed or constructed in the context of a system synthesis or analysis.

In engineering thermodynamics three types of systems are distinguished: open, closed, and isolated systems. In an *open system* both energy and matter can cross the system boundary. Examples of open systems are pumps, turbines, fans, and combustion processes open to the environment. A *closed system* does not allow the passage of matter; only energy can cross its boundaries. Examples of closed systems are electric motors, integrated circuits, mechanical springs, furnaces, and calorimeters. An *isolated system* is completely uninfluenced by its surroundings; it has a fixed mass and no form of energy can cross the boundary. Examples are pumps, electronic circuits, or structural elements that are redundant or which are used in a standby capacity. In some of the literature the closed system is referred to as simply a system and the open system is called a control volume.

The first step in every thermodynamic analysis is to identify the system. This is customarily done by drawing a dotted line that approximates the boundary between the system and its environment.

Figure 1.2 shows an arrangement in which a motor-driven centrifugal pump transports a liquid from one region to another. An integral part of the arrangement is a redundant or standby motor/pump combination which, by definition, is an isolated system. Liquid enters the pump at the suction level and leaves at the discharge level. In addition, work crosses the boundary of the pump, and so it is identified as an open system. In the case of the motor, only electrical and mechanical energy cross the boundary, and so it is a closed system.

1.4 STATE OF A SUBSTANCE

The condition of a substance in a system is referred to as its *state*. If the condition is changed in any way by a transfer of energy, the

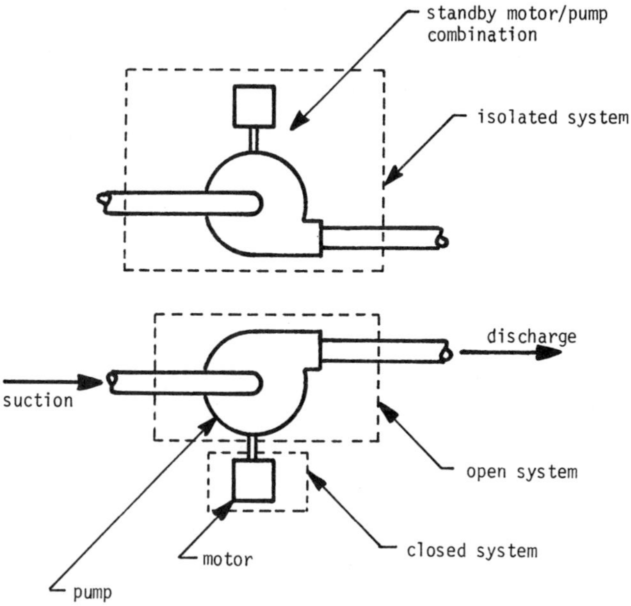

FIGURE 1.2 Schematic of a motor driven centrifugal pump and standby motor/pump combination showing examples of open, closed, and isolated systems.

system is said to have experienced a *change of state*. When a system passes through a succession of states, it goes through a *process*. If a system passes through a succession of processes and returns to the initial state, it has gone through a *cycle*.

A *phase* is a quantity of matter which is homogeneous in physical and chemical composition. The important phases in engineering thermodynamics are solid, liquid, and gaseous; in material science a variety of phases are encountered within the solid condition, but they are not generally considered for thermodynamic analyses. A change in phase such as the transformation from a solid to a liquid or from a liquid to a gas is a special case of a change in state.

A *pure substance* is a quantity of matter that retains the same chemical composition even when it experiences a phase change. Water is a classic example of a pure substance since it has the same chemical identity in the solid, liquid, or gas phase, or in mixtures of these phases. For some thermodynamic analyses a mixture of gases such as air is considered to be a pure substance as long as no phase change occurs. Such an assumption is not strictly true, but the error is small and calculations are greatly simplified as we shall see shortly.

1.5 PROPERTIES OF A SUBSTANCE

A *property* of a substance is a characteristic which can be evaluated quantitatively and is independent of the path through which it reached its value. The state of a substance can be specified by a listing of its properties, and a change in any one of the properties, however small, constitutes a change of state. Examples of properties are mass, volume, temperature, color, and magnetization. Work and heat, which will be defined later, are not properties; they represent transfers of energy that cause changes in properties.

There are two general classifications of properties: intensive and extensive. *Intensive properties* are independent of the size or mass of a system and are not additive. For example, if the size of a system is doubled or halved, all intensive properties remain unchanged. *Extensive properties*, on the other hand, are additive since a doubling of the size of a system doubles all extensive properties. Examples of intensive properties are temperature, pressure, and density, while extensive properties are exemplified by volume, mass, and weight. For many engineering applications, extensive properties are changed to intensive ones by dividing through by a unit mass; examples are specific weight and specific volume. Such a transformation is useful because intensive properties can be considered without specifying particular amounts of a substance.

1.6 EQUATIONS OF STATE

An *equation of state* is a functional relationship between intensive properties of a pure substance. Such an equation is useful because it has been determined experimentally that a pure substance, in the absence of motion, gravity, surface effects, electricity, and magnetism, has only two independent intensive properties. This means that the state of a pure substance is established when two independent properties are specified.

It is possible to conceive of a wide variety of equations of state, but for engineering practice it is customary to use pressure, temperature, and specific volume as the functionally related intensive properties. Equations of state can be written for solids and liquids, but because the intermolecular spacings are so small, a complex system of intermolecular forces exists. As a consequence, each solid and liquid has a unique equation of state and there is no generality. In the case of gases, however, the volume occupied by molecules is small compared with the total volume, and so the intermolecular force field is less complex. Functional relationships between temperature, pressure, and specific volume can be established for gases which are simple enough and general enough to have practical utility.

Equations of state can be presented in several forms: algebraic equations, two- or three-dimensional graphs, or tabulated values. The most widely used algebraic equation is the perfect gas law, which is a simple version of the functional relationship; this law will be covered in some detail in Chap. 8.

Two-Dimensional Graphs

A series of two-dimensional graphs that can be developed experimentally are very useful in thermodynamic analyses. Experimental values can be obtained by placing a substance in a vessel in such a way that pressure, temperature, and specific volume can be monitored. If the substance is a pure substance, the fixing of any two of these intensive properties will establish all other properties of the substance.

For example, consider water in a piston-cylinder arrangement; the water is selected as a closed system into which heat can be added or removed. The arrangement is similar to the one shown in Fig. 1.1 but with the weights removed. The water comprising the system can be in a solid, liquid, or gaseous phase.

Suppose that at the start of the experiment the water is at atmospheric pressure and room temperature; this means it will be in a liquid phase. As the temperature is increased through the addition of heat, the specific volume will increase slightly and the pressure will remain constant. When a temperature of 100°C (212°F) is reached, the liquid will begin to change into a gaseous phase. During the phase change both temperature and pressure will stay constant and the specific volume will increase considerably. After all the water changes from liquid to gas, the temperature and specific volume will increase while the pressure remains constant.

The temperature at which the phase change takes place is called the *saturation temperature* and the corresponding pressure is the *saturation pressure*; in general the water is said to be in a saturated state and is called a saturated liquid. Thus, the saturation temperature for water under a pressure of 1 atm is 100°C (212°C); other pressures would result in different saturation temperatures. Keep in mind that the constant pressure in the experiment described is a result of a freely moving piston. In some other experiments without pistons, different values of pressure and specific volume would result.

The phase formed when a liquid is heated to its saturation temperature is commonly called a *vapor*; the term *gas* is reserved for the condition where the temperature is well beyond the saturation state. *Vaporization* is the term used to describe the phase change from liquid to vapor. Since there is a one-to-one relationship between saturation temperatuıe and saturation pressure, a curve can be plotted from experimental data; such a curve is called a vapor pressure or saturation curve.

Additional data can be obtained if the water is initially in a solid phase at, for example, -273°C (-459°F) and 1 atm pressure. As heat is added to the solid water, the temperature and specific volume increase while the pressure remains constant. When the temperature reaches 0°C (32°F), the water will change to a liquid phase while the temperature and pressure remain constant. At this point a peculiar phenomenon can be noted; the specific volume of the water will decrease. For almost any other substance tested, the specific volume will increase during this process. When all the water has changed to the liquid phase, the temperature and specific volume will increase while the pressure remains constant.

At the point where the phase change begins the water is again said to be in a saturated state, but it is called a *saturated solid*. The change in phase is referred to as *melting* or *fusion*. Obviously, if the temperature is increased still further in this experiment, the vaporization process described above will occur.

If the solid water is initially at -273°C (-459°F), but at a pressure less than 0.006 atm, a different sort of phenomenon will be observed. As heat is added, the water will change directly from solid to vapor without going through a liquid phase. The process is called *sublimation* and, again, it is indicative of a saturation state.

All the experimental results described above can be summarized on a two-dimensional plot of pressure against temperature as shown in Fig. 1.3. The vapor pressure curve is represented as the combination of the sublimation line and the vaporization line; the *fusion* line is the boundary between the solid and liquid phases. Obviously, plotting of the curve requires that a number of different pressures be established in the substance. At two of these pressures, additional phenomena are observed. For example, at 0.006 atm and 0°C (32°F), solid, liquid, and vapor phases coexist in equilibrium; this is called the *triple point* of the substance. Above 218 atm, no sharp distinction between liquid and vapor is observed and this is referred to as the *critical pressure*, which establishes the *critical point*. Table 1.1 shows the triple and critical points for several different substances.

Data from the experiment can be used to plot another useful curve as illustrated in Fig. 1.4. In this graph temperature is plotted against specific volume with pressure as a parameter. The curve is sometimes called a vapor dome and it features the region where a substance exists as part liquid and part vapor.

If a substance exists as a liquid at the saturation temperature and pressure, it is called a saturated liquid; if it exists as vapor

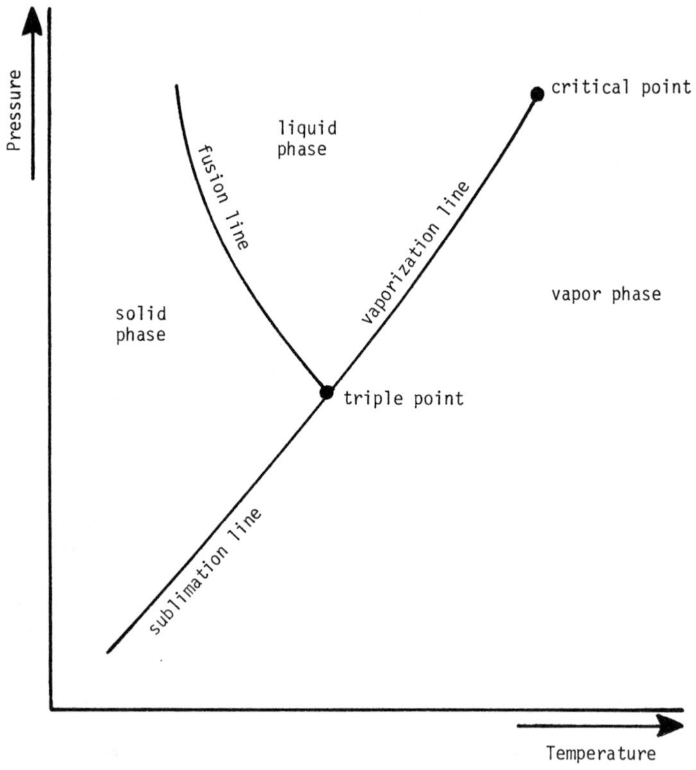

FIGURE 1.3 Pressure-temperature diagram showing the triple point (not to scale).

it is called saturated vapor. Therefore, the saturation temperature and pressure for both conditions are the same. However, the saturated liquid and saturated vapor are very definitely not the same state. This poses a problem since two intensive properties of a pure substance identify two different states. This problem is resolved by realizing that temperature and pressure are *not* independent properties in a saturation state.

Consider the behavior of the water system in Figure 1.4 when the pressure is maintained at 1 atm. Assume that point a is at room temperature and point b is at 100°C (212°F). In this region the

TABLE 1.1 Critical Constants and Triple Point Data for Several Substances

Substance		Critical Point Date				Triple Point Date			
		Temperature		Pressure		Temperature		Pressure	
WATER	H_2O	1165.3	°R	3208	psi	491.69	°R	0.0886	psi
		647.4	K	22.118	MPa	273.16	K	610.6	Pa
HELIUM	H_2	59.9	°R	188.1	psi	24.91	°R	1.02	psi
		33.3	K	1.297	MPa	13.84	K	7039.4	Pa
NITROGEN	N_2	227.1	°R	492	psi	113.72	°R	1.82	psi
		126.2	K	3.392	MPa	63.18	K	12532.3	Pa
OXYGEN	O_2	278.6	°R	736	psi	97.85	°R	0.022	psi
		154.8	K	5.075	Ma	54.36	K	152.0	Pa

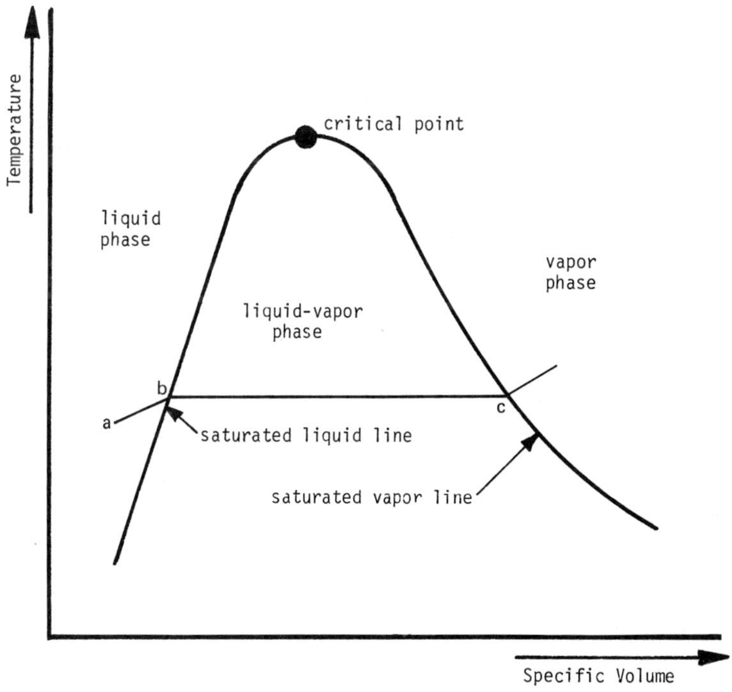

FIGURE 1.4 Temperature-specific volume for water (not to scale).

water is below the saturation temperature for the existing pressure (or above the saturation pressure for the existing temperature) and is referred to as a *subcooled liquid* or a *compressed liquid*.

Region bc under the vapor dome shows the relatively large increase in specific volume of the water during the vaporization process while the temperature and pressure remain constant. Beyond point c the substance is said to be a *superheated vapor* and the temperature again increases. When the temperature is increased to some level, not definitely established, the substance ceases to be a vapor and becomes a gas.

Under the vapor dome the substance is said to have an *intensive property* called quality, x:

$$x \equiv \frac{\text{mass of vapor}}{\text{total mass of substance}} \tag{1.1}$$

Note that saturated liquid has 0% quality while saturated vapor has 100% quality. The latter is sometimes referred to as *dry saturated vapor*. The concept of quality has meaning only when a substance is in a saturated state and is at the saturation temperature and pressure. Also, as previously stated, it is obvious that temperature and pressure are *not* independent in the saturated state.

Three-Dimensional Thermodynamic Surfaces

As the temperature and pressure of a system are varied, a substance comprising the system changes phase from solid to liquid to vapor. A useful illustration of this relationship is obtained when a three-dimensional surface is constructed where temperature, pressure, and specific volume are plotted on three mutually perpendicular axes. Two models are necessary: one representing a substance, such as water, which expands when it freezes, and one representing most other substances which contract upon freezing. Figures 1.5 and 1.6, respectively, illustrate the two models.

Line abcdef on each thermodynamic surface shows the state of a substance as heat is added and the pressure is kept constant. Initially the substance is a solid at some temperature, pressure, and

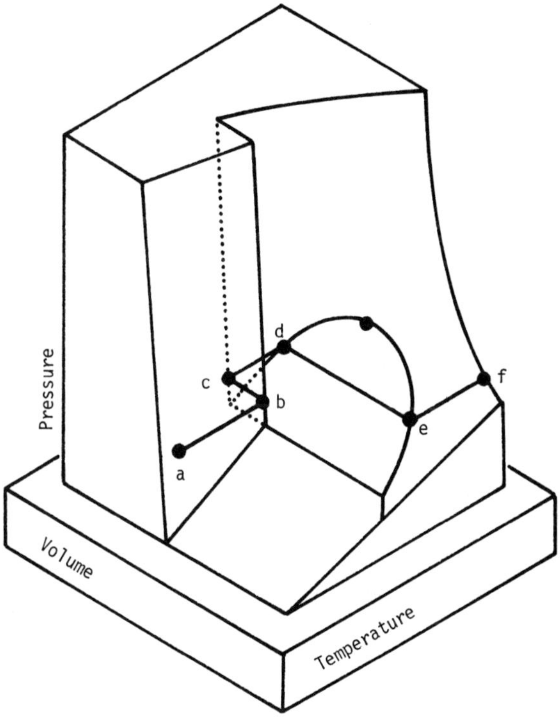

FIGURE 1.5 PVT surface for a pure substance (such as water) which expands when freezing. Isobaric process abcdef is shown (not to scale).

specific volume indicated by point a. As heat is added the temperature of the substance is raised in a manner determined by the specific heat, and the volume is increased slightly according to the thermal coefficient of expansion. At point b on the surface, melting of the substance begins and it changes from solid to liquid; temperature remains constant even though heat is still being added to the substance. In Fig. 1.5 (for water) specific volume decreases, while in Fig. 1.6 specific volume increases. The substance has changed completely to liquid at point c in both models.

The continued addition of heat increases the liquid temperature according to its specific heat, and increases the specific volume according to the thermal coefficient of expansion. At point d the

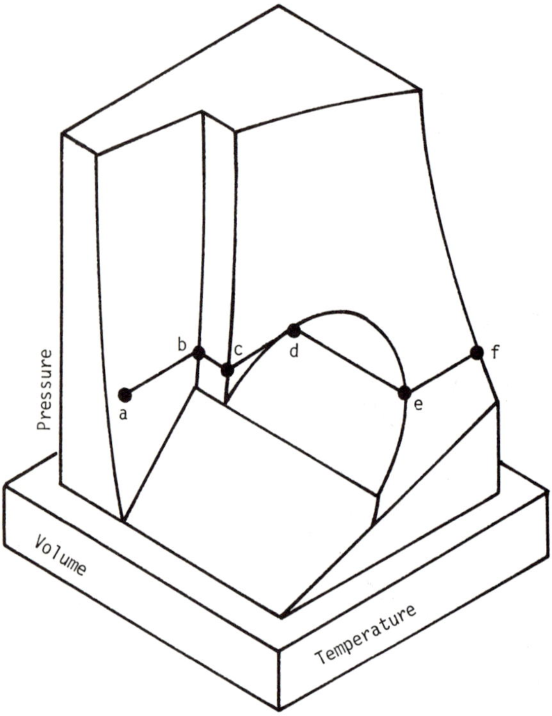

FIGURE 1.6 PVT surface for a pure substance which contracts when freezing. Isobaric process abcdef is shown (not to scale).

substance is a saturated liquid, and line de traces the vaporization process to point e where the substance is a saturated vapor. During this process the substance is a two-phase fluid and the heat capacity and thermal coefficient of expansion for both the liquid and vapor must be considered. Line ef traces the path into the superheated region.

If a heat sink replaces the heat source, it is possible to transfer heat from the substance to the environment. For such a system the processes of Figs. 1.5 and 1.6 can be reversed. Also, it is obvious that the two-dimensional graphs described in the previous section can be obtained by cutting the three-dimensional thermodynamic surfaces with appropriate planes.

Thermodynamic Property Tables

Over the years a considerable effort has been expended by many researchers to measure thermodynamic properties for a wide variety of substances. These data have been tabulated and plotted in different forms and publications of these works have greatly simplified thermodynamic analyses for engineers and technologists. The reference list at the end of this chapter includes a number of these publications [1-5].

The *Steam Tables* by Keenan, Keyes, Hill, and Moore [1,2] are probably the most widely known thermodynamic tables, and an explanation of them will make all other tables understandable. As a convenience, the first of the steam tables is in English units and the second is in metric units.

Tables 1 and 2 from the steam tables present the same data for water in a saturated state; the only difference is that table 1 lists the data as a function of saturation temperature while table 2 uses saturation pressure. The first column after the saturation temperature in table 1 is the corresponding saturation pressure; in table 2 the saturation temperature is given for the listed saturation pressure. Both tables give thermodynamic values of four properties: specific volume, internal energy, enthalpy, and entropy. The latter three properties will be defined and discussed in subsequent sections.

Subscript f in the tables refers to a property of liquid in equilibrium with vapor, subscript g refers to a property of vapor in equilibrium with liquid, and fg refers to a change by evaporation. It follows from this designation that

$$v_g - v_f = v_{fg} \tag{1.2}$$

This equation, used with the definition of quality, allows us to determine the specific volume of a substance at any point between saturated liquid and saturated vapor. For any unit mass of a substance, the volume of the vapor is xv_g and the volume of the liquid is $(1 - x)v_f$. Also, the specific volume is the sum of these two quantities:

$$v = xv_g + (1 - x)v_f \tag{1.3}$$

$$v = v_f + xv_{fg} \tag{1.4}$$

$$v = v_g - (1 - x)v_{fg} \tag{1.5}$$

Similar equations can be written for the other properties in the steam tables and they will be shown when the properties are introduced.

Tables 3 and 4 list the thermodynamic properties as a function of temperature and pressure for a superheated vapor and a subcooled liquid, respectively. Remember that in these unsaturated states temperature and pressure are independent intensive properties.

1.7 CONCLUSION

This chapter presents some fundamental definitions and concepts of thermodynamics. Equilibrium and quasiequilibrium processes are defined from a continuum viewpoint, and the system concept for thermodynamic analysis is described. States of substances and their dependence on properties are discussed, and relationships of state on processes and cycles are pointed out. Equations of state as a consequence of the concept of pure substances are presented in several forms.

Topics introduced in this chapter will be referred to in subsequent chapters as further developments are made. The serious student of thermal technology would do well to be acquainted with coverage of the same material by other authors [6-8].

SOLVED PROBLEMS

PROBLEM 1: Determine the specific volume for steam of 60% quality at 425°F using (a) Eq. (1.3), (b) Eq. (1.4), and (c) Eq. (1.5).

Solution: From table 1 of Keenan and Keyes *Steam Tables* [1], at 425°F:

$$v_f = 0.019014 \text{ ft}^3/\text{lbm}$$

$$v_g = 1.4249 \text{ ft}^3/\text{lbm}$$

from Eq. (1.2),

$$v_{fg} = v_g - v_f = 1.4249 - 0.019014 = 1.40589 \text{ ft}^3/\text{lbm}$$

(a) From Eq. (1.3),

$$v = xv_g + (1 - x)v_f$$
$$= (0.6)(1.4249) + (0.4)(0.019014) = 0.86254 \text{ ft}^3/\text{lbm}$$

(b) From Eq. (1.4)

$$v = v_f + v_{fg}$$
$$= 0.019014 + (0.6)(1.40589) = 0.86254 \text{ ft}^3/\text{lbm}$$

(c) From Eq. (1.5)

$$v = v_g - (1 - x)v_{fg}$$
$$= 1.4249 - (0.4)(1.40589) = 0.86254 \text{ ft}^3/\text{lbm}$$

PROBLEM 2: Saturated steam at 350 lbf/in^2 is heated to 500°F. Sketch a temperature-volume vapor dome and show the specific volume, temperature, and pressure for the saturated and superheated states.

Solution: From Keenan and Keyes *Steam Tables* [1]:
At 350 lbf/in^2, table 2, saturation pressures,

$$t_{sat} = 431.82°\text{F}$$
$$v_f = 0.019124 \text{ ft}^3/\text{lbm}$$
$$v_g = 1.3267 \text{ ft}^3/\text{lbm}$$

At 500°F and 350 psia, table 3, vapor,

$$v = 1.4913 \text{ ft}^3/\text{lbm}$$

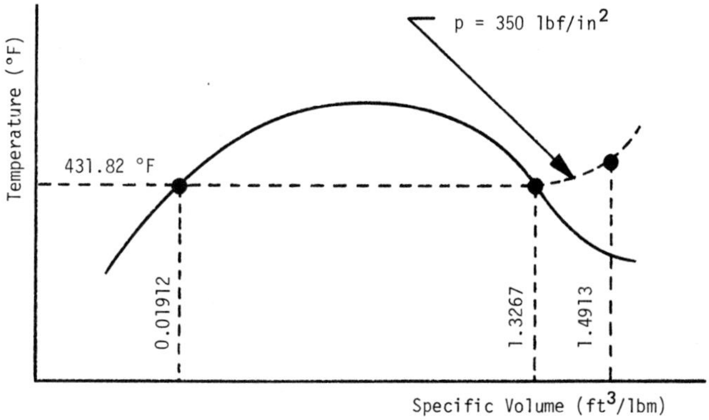

PROBLEM 3: Table 4 of the steam tables shows the slight compressibility of water. If saturated liquid steam at 100°C has its pressure increased to 200 bar while the temperature remains constant, determine how much the liquid is compressed.

Solution: From Keenan and Keyes *Steam Tables* [2]:
At 100°C, table 1, saturation temperatures,

$$p_{sat} = 1.0135 \text{ bar}$$

$$v_f = 1.0435 \text{ cm}^3/\text{gm}$$

At 200 bars and 100°C, table 4, liquid,

$$v = 1.0337 \text{ cm}^3/\text{gm}$$

∴ liquid is compressed $1.0435 - 1.0337 = 0.0098 \text{ cm}^3/\text{gm}$

EXERCISES

1. Give an example of (a) a closed system and (b) an open system. Make a schematic drawing of each and define the environment and the boundary.
2. Give an example of a system that includes at least one subsystem and two or more components. Under what circumstances would the components be considered as systems?
3. Give an example of a cycle. Describe the processes that make up the cycle and indicate several changes of state that occur.
4. Indicate whether the following properties are intensive or extensive: (a) specific weight, (b) specific gravity, (c) molecular weight, (d) color, and (e) thermal conductivity.
5. Sketch a temperature-volume vapor dome for nitrogen. Identify the critical point and show a line of constant pressure in the two-phase area of the dome and in the superheated area.
6. Identify a practical example of (a) evaporation, (b) boiling, (c) freezing, and (d) sublimation.
7. The specific volume of water increases as it is cooled at temperatures near the freezing point. What would be the unfortunate result if this phenomenon were reversed?

8. Determine by the use of published steam tables the specific volume for (a) saturated vapor at 180°F, (b) superheated vapor at 180°F and 7.5 psia, and (c) subcooled liquid at 50°F and 1500 psia.
9. Using metric units, calculate the specific volume of saturated steam at 160°C and 15% quality from (a) Eq. (1.3) and (b) Eq. (1.5).
10. Saturated steam at a pressure of 7 bars is heated to 250°C. Sketch a temperature-volume vapor dome and show the specific volume, temperature, and pressure for the saturated and superheated states.
11. Determine the change in specific volume of saturated liquid steam at 200°C if the pressure is increased to (a) 400 bars and (b) 25 bars. What conclusion does the results suggest?

REFERENCES

1. J. H. Keenan, F. G. Keyes, P. G. Hill, and J. G. Moore, *Steam Tables: English Units*, Wiley, New York, 1969.
2. J. H. Keenan, F. G. Keyes, P. G. Hill, and J. G. Moore, *Steam Tables: Metric Units*, Wiley, New York, 1969.
3. J. H. Keenan and J. Kaye, *Gas Tables*, Wiley, New York, 1948.
4. *Refrigerant Tables, Charts, and Characteristics*, ASHRAE, 1959.
5. R. C. Weast (ed.), *Handbook of Physics and Chemistry*, 49th ed., pp. E12-E27, Chemical Rubber Co., Chicago, 1969.
6. G. J. Van Wylen, *Thermodynamics*, pp. 13-46, Wiley, New York, 1959.
7. W. C. Reynolds, *Thermodynamics*, pp. 1-21, McGraw-Hill, New York, 1965.
8. F. W. Sears, *Thermodynamics*, pp. 1-25, Addison-Wesley, Reading, Mass., 1953.

2

UNITS AND DIMENSIONS

2.1 INTRODUCTION

Mathematics is the language or symbolism in which physical laws are expressed so that the meaning of the laws is coherent and unambiguous. The elements of mathematics are numbers, which have no relation to physical laws unless the numbers can be associated with units and dimensions. It follows, then, that units and dimensions are the connecting links between the laws of nature and the equations of mathematics.

Much of the difficulty encountered in engineering calculations is due to the many physical quantities which must be expressed symbolically and then combined mathematically. The difficulty is aggravated by the multiplicity of systems which have been established for units and dimensions. In the sections that follow, seven distinct systems will be described and most of them have current usage. Different disciplines have jealously nurtured systems of their choice, and so it has been difficult to obtain universal agreement on any particular system of units and dimensions.

Practically all of the nations in the world have officially adopted the International Metric System, or will do so in the near future. However, a significant amount of literature still exists in which other systems of units and dimensions are used. This means that a serious practitioner of science and technology will find it necessary to be familiar with a variety of systems for an indefinite period of time.

2.2 NATURE OF UNITS AND DIMENSIONS

Since any system of units and dimensions is completely arbitrary, it is essential that careful definitions and rules be established. We begin by defining the elements themselves.

A *dimension* is defined as the qualitative measure of a physical property; only one dimension is assigned to each property. Examples of dimensions are length, time, velocity, and temperature.

A *unit* is defined as the quantitative measure of a physical property; a variety of units may be assigned to the same property. For example, length may be expressed in feet, centimeters, angstroms, or light-years.

There are two classes of dimensions used in scientific and engineering calculations: *basic* and *derived*.

A *basic dimension* (primary dimension) is one which may be measured directly and which is independent of other dimensions. From a mathematical viewpoint, a basic dimension is expressed as a first power, e.g., length is expressed as L^1, while area, which is length × length, is L^2, and volume is L^3. Other basic dimensions are time T, force F, mass M, temperature θ, heat H, and electric current I.

Basic dimensions are often determined by the measuring techniques available, or sometimes by the uses to which the dimensions are put. Length is basic to most systems because it is convenient to measure and because it is an element of geometry. It is conceivable, however, that area (L^2 or A) could be considered as a basic dimension; this is actually the case for the engineer who uses a planimeter for determining the area under a curve. In such an instance, length would not be expressed as a first power but rather $A^{1/2}$. Another similar example is velocity, which is a derived dimension in most cases; however, the traffic policeman monitoring a radar speed trap reads the velocity (or speed) directly on his scope, and so he considers it a basic dimension.

A *derived dimension* (secondary dimension) is one which usually cannot be measured directly, and which depends upon other dimensions for its definition. This dependence is characterized by something called "dimensional representation," illustrated by the following examples:

$$\text{Stress} \equiv \frac{\text{force dimension}}{\text{area dimension}}$$

$$\text{Area} \equiv \text{length dimension} \times \text{length dimension}$$

Dimensions are abstract concepts which have meaning because they are defined in a particular fashion, usually based on the existence of some natural law. For example, the concept of *length* exists because in our world we observe that two bodies cannot exist in the same location concurrently. The dimension of length allows us to differentiate one location from another, i.e., we say that body A is a certain length from body B. The concept of *time* may be deduced by the further observation that bodies A and B could occupy the same location if they were placed there sequentially. The dimension that measures the separation of the sequences is called time.

The dimensions of length and time are fundamental, as evidenced by the fact that they were apparent to the most primitive men and are sensed by members of the animal world, to whom we do not usually attribute any significant reasoning power. As the degree of abstraction is increased in the definition of dimensions, a corresponding increase in intelligence is required to comprehend them. This is particularly true for derived dimensions which are defined in terms of basic dimensions, e.g.,

$$\text{Velocity} \equiv \frac{\text{length dimension}}{\text{time dimension}}$$

The lion, deciding whether or not he can spring upon an antelope, is quite aware that a distance separates him from his prey and that a certain period of time will be required for him to reduce the distance to zero. It is most unlikely, however, that the lion thinks in terms of a velocity dimension.

Proceeding up the scale of abstraction, we may consider the dimension of thermal conductivity. This dimension is defined as the amount of heat which flows through a unit area in a unit time when the temperature gradient is unity. It is evident that such a concept would have been incomprehensible to any primitive man; indeed, it can even be abstruse for a modern man unless he has some training in thermoscience.

The determination of which dimensions are basic and which are derived is completely arbitrary. Different branches of science and technology have found it convenient to work with different systems, and these systems will be described in later sections.

When a dimension is expressed in a particular unit, an indication is obtained of how much of the physical property is under consideration. Each unit is a relative quantity and must include three considerations:

1. A datum
2. An operational procedure to define the unit
3. An operational procedure for interpolation and extrapolation

These considerations are best explained by considering an example of a quantitative measure of length, such as the inch:

1. The datum is zero.
2. The operational procedure for definition is comparison against a standard inch.
3. Interpolation and extrapolation is obtained through a system of fractions and decimals of an inch.

The above example is simple and straightforward, but this is not always the case. The units for dimensions such as enthalpy, entropy, and temperature do not have zero for a datum, but rather, some arbitrarily defined point. In addition, the units are not defined by a simple comparison against a standard. This point can be illustrated by considering the types of units which are used to measure temperature:

1. The datum is not zero, but rather, the triple point of water.

2. The operational procedure for defining the unit depends on the method of measurement. For example, temperature is determined by the change in some physical property such as a thermal expansion of solids, liquids, and gases; electrical resistance; or surface color. The ordinary household thermometer works on the principle that mercury expands about the same amount for each degree of temperature rise over the range -30°F to +1000°F. It is evident then that mercury thermometers should not be used for extremely high- or low-temperature measurements. In another application, the blacksmith will observe the color changes in heated metal in order to estimate the temperature before quenching. A method of this type is obviously limited in accuracy.

2.3 CONVERSION OF UNITS AND DIMENSIONS

If mathematical techniques are to be applied to units and dimensions, there must be a consistent and convenient symbolism established. In the case of units, it is universally agreed that equations of the following type may be written:

$$1 \text{ ft} = 12 \text{ in}$$

or

$$\frac{1 \text{ ft}}{12 \text{ in}} = 1.0$$

Such symbolism permits one unit to be converted to another unit in an orderly fashion. As an example, 10 ft can be expressed in terms of inches in the following manner:

$$10 \text{ ft} = 10 \text{ ft} \frac{12 \text{ in}}{\text{ft}} = 120 \text{ in}$$

Granted, this example is trivial. But many errors can result in converting some of the more complex units if the orderly method shown above is not followed.

Symbolism for dimensions is not entirely consistent, but in general, capital letters are used to designate basic dimensions: L for length, T for time, M for mass, F for force. Square brackets are often used to impart the meaning "the dimension of" or "the unit of." As an example, velocity, which is defined as the ratio of length to time, can be expressed dimensionally as

$$[\text{Velocity}] = [\frac{L}{T}]$$

and read as "the dimension of velocity equals the dimension of length over the dimension of time." Velocity can also be expressed in terms of units:

$$[\text{Velocity}] = [\frac{m}{s}]$$

and read as "the units of velocity equal the units of meters per second."

It is possible to cancel and convert dimensions in the same manner that is used for units. For example, let us consider the well-known relation that expresses velocity as the product of acceleration and time, and check the dimensions of velocity from the equation

$$[\text{Velocity}] = [\frac{L}{T}]$$

$$[\text{Acceleration}] = \left[\frac{L}{T^2}\right]$$

$$[\text{Time}] = [T]$$

$$\therefore \quad [\frac{L}{T}] = \left[\frac{L}{T^2}\right][T] = [\frac{L}{T}] \quad \text{(check)}$$

Once again, the reader is cautioned not to be deceived by the simplicity of the example. The cancellation procedure shown can prevent many errors when equations are complex and many different dimensions are involved [1].

2.4 GENERAL CLASSIFICATIONS

One system of units and dimensions would be sufficient if everyone agreed on a standard meaning for each physical variable and everyone recognized the same basic units and dimensions. Such agreement will result if everyone adopts the International Metric System, but at this time such agreement does not exist. As a consequence, several systems are in common usage. Fortunately there is a connecting link between the systems, namely, Newton's second law of motion, which is the basis of them all. The relationship of this law will be expanded in a later section.

There are two general classifications of systems, the *metric* and the *British*, which differ in the types of units used. Under these classifications may be found the absolute and gravitational systems. An *absolute system* is arbitrarily defined as one in which length, mass, and time are selected as basic dimensions and a *gravitational system* as one in which length, force, and time are basic dimensions. In addition there are subclassifications such as "practical," "technical," or "engineering," which modify the general systems in one way or another.

Much of the confusion encountered in working with units and dimensions results from the distinction between force and mass. This is particularly true in the British systems where the term "pound" is used to express both quantities. A prudent rule to observe is never to use the term "pound" or the symbol "lb" but, instead, "pound of mass" or "pound of force," lbm and lbf, respectively.

2.5 NONDIMENSIONAL QUANTITIES

There is a small number of quantities which are nondimensional in nature, i.e., the exponent of the dimension is zero. Such quantities are ratios of physical quantities of the same type. An example is specific gravity, which is the ratio of the density of a substance to the density of water. Similarly, a radian, defined as the central angle subtended by a length of circular arc equal to the radius of the circle, has a dimension of $L/L = L^0$. Another example is efficiency, defined as the ratio of output to input of either power or energy.

It is a common practice to plot various types of nondimensional quantities in order to analyze physical systems. Examples of one type of these quantities are the nondimensional numbers used in Dimensional Analysis. Moduli such as the Reynolds, Prandtl, and Nusselt numbers are obtained by combining physical properties in such a way that all dimensions cancel. Other types of nondimensional quantities are formed from the ratio of the value of some property to the value at a selected reference point. Some examples are the density of a gas divided by its density under standard atmospheric conditions, or the temperature of an extended surface divided by the temperature at the base. When nondimensional quantities are plotted, the resulting curves have universal validity and can be used to compare the relative behaviors of systems of different size.

When mathematical transformations such as Laplace and Fourier transforms, logarithms, or trigonometric and hyperbolic functions are used to solve equations, no units or dimensions are associated with the numbers. Great care should be taken, however, to make sure that the equations so solved are dimensionally homogeneous before the transforms are applied.

2.6 SIMILAR DIMENSIONS APPLIED TO DISSIMILAR QUANTITIES

Dimensional formulas cannot account for the sense in which a quantity is directed. As a consequence, certain physically different quantities have the same dimensions. The classic example is the comparison of work and torque. Work is defined as the product of force and distance, where the force is in the direction of the displacement; the dimension of work is FL. Torque has the same dimensions, but the force is directed at right angles to the length. Another striking example is the section modulus for a beam. This modulus is specified as the ratio of the area moment of inertia (dimension of L^4) and the distance from the neutral axis (dimension L). The resulting dimension of the section modulus is $L^4/L = L^3$, or the same as volume. Such peculiarities should not cause trouble as long as the physical significance of the quantities is kept in mind.

2.7 SYSTEMS OF UNITS

The first effort to develop a consistent system of units was undertaken by the French Academy of Sciences in 1790. After a period of study the Academy proposed the adoption of the metric system, and the proposal was approved by the French Government in 1795. The system as proposed was absolute, that is, mass was chosen as a basic dimension with force a derived dimension. Through a misinterpretation, some engineers converted the system into a gravitational one by using the weight of the standard mass as a basic dimension. In 1881 another metric system, the cgs absolute system, was adopted by the first International Electrical Congress, and subsequently became firmly established in many branches of science. While these developments were underway, several British systems, both absolute and gravitational, were being used widely in English-speaking countries. The multiplicity of usage has continued to the present day so that a number of different systems are now "standard," depending on the type of discipline being considered.

All countries in the world having committees concerned with science and technology standards approved the adoption of an international system in 1954. The action was taken at the 10th General Conference on Weights and Measures where the International MKSA system of units was formally approved. This system is almost identical to the metric system proposed by the French Academy of Sciences in the latter part of the eighteenth century.

Unfortunately, many of the countries that approved the International MKSA system did not take strong action to enforce, or at least encourage, its use. Numerous papers and books being published today (particularly in English-speaking countries) continue to use the cgs or British systems. It is unlikely, therefore, that these systems will be replaced for some time. The only country that has taken positive action is France, where after January 1, 1962 it became a punishable offense to use in contracts or official documents, or on instruments to be sold, any units other than those specified by the International MKSA system.

Because of the inertia encountered in the adoption of a universally standard system of units and dimensions, a wide variety of systems are found in even the modern literature. As a consequence, the student and practitioner of science and technology must be knowledgeable in a number of different systems. In the sections that follow, some of the most widely recognized systems are described by means of a listing of their basic units and dimensions. Table 2.1 is a summary of the fundamental quantities of each system.

International System of Units (SI Units)

The International System of Units had its beginnings during the French Revolution. It was based primarily on the meter, a length defined as a small portion of the earth's circumference. Since then the system has been revised and extended. Hopefully, at some time in the future, it will be the only system of units and dimensions in use throughout the world. The official name is the Systeme International d'Unites, and in this country the units are called SI units.

The basic units (see Table 2.2) in this absolute system are the meter (m), kilogram (kg), second (s), ampere (A), Kelvin (K), and candela (cd). Force is a derived dimension and its unit, the newton (N), is defined by Newton's second law of motion as

$$\text{Force} = \text{mass} \times \text{acceleration}$$

$$[\text{Force}] = [\text{kg}][\text{m/s}^2] = \text{newton}$$

This system has been deemed worthy for universal adoption because it is consistently applicable to mechanical, thermal, electrical, and illumination phenomena. Complete listings of the units included in this system have been published by the International Organization for Standardization [2].

Metric Absolute System (cgs Units)

The basic units in the metric absolute system are the centimeter (cm), gram (g), and second (s). Force is a derived dimension and its unit, the dyne (dyn), is defined as

TABLE 2.1 Units and Dimensions of the Major Systems

	System	Metric						British							
		Absolute				Gravitational		Absolute		Gravitational				Engineering	
		Dim	Unit	Dim	Unit	Dim	Unit	Dim	Unit	Dim	Unit	Dim	Unit	Dim	Unit
Basic Dimensions & Units		L	cm	L	m	L	m	L	ft	L	ft	L	ft	L	ft
		M	gm	M	kg	F	kgf	M	lbm	F	lbf	F	lbf	M	lbm
		T	sec	T	sec	T	sec	T	sec	T	sec	T	sec	F	lbf
		-	-	-	-	-	-	-	-	-	-	-	-	T	sec
Derived Dimensions	FORCE	dyne		newton		-		poundal		-		-		-	
	MASS	-		-		metric slug		-		slug		$mass = \frac{W}{g}$		-	
	WORK	dyne-cm or erg		newton-meter or joule		kgf-m		ft-poundal		ft-lbf		ft-lbf		ft-lbf	
	POWER	$\frac{erg}{sec}$		joule/sec or watt		$\frac{kgf\text{-}m}{sec}$		$\frac{ft\text{-}poundal}{sec}$		$\frac{ft\text{-}lbf}{sec}$		$\frac{ft\text{-}lbf}{sec}$		$\frac{ft\text{-}lbf}{sec}$	

TABLE 2.2 Fundamental Units and Dimensions of the International Metric System

Dimension	Dimensional Symbol	Unit	Unit Symbol
Length	L	meter	m
Mass	M	kilogram	kg
Time	T	second	s
Temperature	θ	degree Kelvin	K
Electric Current	I	ampere	A
Luminous Intensity	not resolved	candela	cd

$$[\text{Force}] = [\text{g}][\text{cm/s}^2] = \text{dyn}$$

This system was devised at the initiative of the British Association for the Advancement of Science in 1881. It was based on the premise that the unit of length should be that which made the mass of a cube of pure water at standard pressure and 4°C equal to the unit mass of a gram. The system is used widely by physicists and appears in much of the literature of pure and applied science.

British Absolute System

The basic units for the British absolute system are the foot (ft), the pound mass (lbm), and the second (s). Force is a derived dimension and its unit, the poundal, is defined as

$$[\text{Force}] = [\text{lbm}][\text{ft/s}^2] = \text{poundal}$$

This system is used in some of the older elementary textbooks, but is not widely used in modern practice.

British Gravitational System

The basic units for the British gravitational system are the foot (ft), the pound force (lbf), and the second (s). Mass is a derived dimension and its unit, the slug, is defined as

$$[\text{Mass}] = [\text{lbf}][\text{s}^2/\text{ft}] = \text{slug}$$

This system attempts to overcome a disadvantage of the British absolute system by shifting the unwelcome burden of an unfamiliar unit from the force to the mass dimension. This is the system which authors of physics textbooks like to attribute to engineers, but which in practice is used by few engineers.

Modified British Gravitational System

The modified British Gravitational system was devised to avoid the use of the slug as a mass unit. It does so by not defining mass, per se, but by using the concept of weight instead. Weight W is defined as the force exerted by gravity g. It can be expressed in terms of mass by using Newton's second law where the force becomes the weight and the acceleration becomes the acceleration of gravity:

$$W = Mg$$

$$\therefore \quad [\text{Mass}] = [\text{lbf}][\text{s}^2/\text{ft}]$$

Therefore, in this system the basic units are the foot (ft), the pound force (lbf), and the second (s). The dimension of mass is not used since it is replaced by W/g. Newton's second law then takes the following form:

$$F = \left(\frac{W}{g}\right)(a)$$

$$[\text{lbf}] = [\text{lbf}][\text{s}^2/\text{ft}][\text{ft}/\text{s}^2] = [\text{lbf}]$$

This system is widely used by engineers and some scientists who wish to use British units. It is found most extensively in the literature devoted to dynamics, statics, vibrations, and strength of materials.

Metric Gravitational System

The basic units for the metric gravitational system are the meter (m), kilogram force (kgf), and the second (s). Mass is a derived unit and its unit, the metric slug, is defined as

$$[\text{Mass}] = [\text{kgf}][\text{s}^2/\text{m}] = \text{metric slug}$$

This system has been used by some European engineers but is now being replaced by the international metric system. Where it is still in use, the unit kilopond (kp) is often ascribed to the force dimension, to distinguish it from the kilogram mass.

British Engineering System

In the British engineering system four basic dimensions are used instead of three, namely, length, mass, force, and time. The units are the foot, the pound mass, the pound force, and the second. Because the expression "pound" is used for both mass and force, the gravitational constant must differ from unity. The magnitude and units of this constant can be evaluated by considering Newton's law which states that the force on a body of fixed mass is proportional to the product of mass and acceleration:

$$F \propto Ma$$

In order to write the expression as an equality, a proportionality constant must be included which can arbitrarily be designated as $1/g_c$:

$$\therefore \qquad F = \frac{Ma}{g_c}$$

But a standard gravitational field of 32.2 ft/s^2 will exert one pound of force on a one pound mass, i.e.,

$$1 \text{ lbf} = \frac{1 \text{ lbm}}{g_c} 32.2 \text{ ft/s}^2$$

$$\therefore \qquad g_c = 32.2 \text{ lbm ft/lbf s}^2$$

This value of the gravitational constant correlates the units of the British engineering system. It is constant even in the face of a variable gravitational field. For example, one can calculate the force due to gravity on a 1 lbm when the acceleration of gravity is 30 ft/s^2:

$$F = \frac{Ma}{g_c}$$

$$= \frac{(1 \text{ lbm})(30 \text{ ft/s}^2)}{(32.2 \text{ lbm ft/lbf s}^2)} = 0.93 \text{ lbf}$$

In other words, this is the weight of 1 lbm in such a gravitational field.

All the systems of units discussed thus far have corresponding gravitational constants associated with them. They are not used in practice because their value is unity, although they do have units. As an example, consider g_c in the cgs system:

$$F = \frac{Ma}{g_c}$$

$$1 \text{ dyn} = \frac{(1 \text{ gm})(1 \text{ cm/s}^2)}{g_c}$$

$$g_c = 1 \frac{\text{gm cm}}{\text{dyn s}^2}$$

The British engineering system is widely used by engineers in the fields of thermodynamics, fluid mechanics, and heat transfer.

Hybrid Systems

The basic units and dimensions discussed above are summarized in Table 2.1. In practice, however, scientists and technologists often use hybrid versions of the pure systems. For example, the thermal engineer prefers a time unit of an hour because significant amounts of heat are not normally transferred in a second. Conversely, the electronic circuit designer is concerned with very small time intervals, and so he works with time units of nanoseconds or picoseconds. The physical chemist expresses length dimensions in angstroms, while the astronomer uses light-years or megaparsecs (3.26×10^6 light-years). The practice of using latin and greek prefixes to increase or decrease a unit has been standardized [3] and a listing of these prefixes is given in Table 2.3.

The commonly used electrical units such as the volt, ohm, and watt are defined in the International MKSA system, but they are also used in other systems. Examples of this usage are the units watts per square inch and ohm-centimeter. The occurrence of hybrid units should not be troublesome, but anyone doing interdisciplinary study or reference can be confused by the practice of the same symbol being used to represent different quantities. It is not uncommon for the letter H to represent heat, enthalpy, and magnetic field strength; the letter I to designate electric current and two kinds of moment of inertia; or to have J stand for current density, the mechanical equivalent of

TABLE 2.3 Mathematical Prefixes for Units

Multiplier			Prefix	Symbol
1 000 000 000 000 000 000	=	10^{18}	exa	E
1 000 000 000 000 000	=	10^{15}	peta	P
1 000 000 000 000	=	10^{12}	tera	T
1 000 000 000	=	10^{9}	giga	G
1 000 000	=	10^{6}	mega	M
1 000	=	10^{3}	kilo	K
100	=	10^{2}	hecto	h
10	=	10^{1}	deka	da
Base unit 1	=	10^{0}		
0.1	=	10^{-1}	deci	d
0.01	=	10^{-2}	centi	c
0.001	=	10^{-3}	milli	m
0.000 001	=	10^{-6}	micro	μ
0.000 000 001	=	10^{-9}	nano	n
0.000 000 000 001	=	10^{-12}	pico	p
0.000 000 000 000 001	=	10^{-15}	femto	f
0.000 000 000 000 000 001	=	10^{-18}	atto	a

heat, and polar moment of inertia. Fortunately, these diverse quantities are usually not considered concurrently when some particular phenomenon is being investigated.

2.8 DERIVED DIMENSIONS

Three basic dimensions (length, force or mass, and time) are sufficient for developing all derived dimensions used for mechanical, thermal, and photometric quantities. In thermodynamics the basic dimension of temperature θ is introduced, but it is strictly a matter of convenience since temperature can be expressed in terms of other established dimensions by use of an equation of state such as the perfect gas law.

Electrical quantities, however, are distinct and unique, so that a fourth basic dimension is necessary. This fourth dimension has been variously selected as electric charge, resistance, and electric current; the latter quantity is specified as the basic electrical dimension in the International System of Units.

Derived dimensions are developed from physical laws or from definitions of quantities. Consider, for example, the classical definition of work as a force being moved through some displacement:

$$[\text{Work}] = [\text{FL}]$$

In the SI system the unit of work is the joule, which is one newton-meter, and in the British system it is the foot-pound force. Time is not considered in the concept of work. The same amount of work is required to raise a weight no matter how long it takes. Since the time rate of work is obviously important in many problems, the concept of power is available:

$$[\text{Power}] = \frac{[\text{work}]}{[\text{time}]} = \left[\frac{\text{FL}}{\text{T}}\right]$$

The unit of power in the SI system is the joule per second or the watt. The ft lbf/s is such a small unit that the horsepower (hp) is substituted for it:

$$1 \text{ hp} = 550 \text{ ft lbf/s}$$

It has been customary in the United States to express mechanical power in horsepower and electrical power in watts. Such a distinction, however, is not necessary. It is possible to express the dissipation of a transistor in horsepower and the power of an automobile in watts.

Many other derived dimensions have been developed for various kinds of analyses. Some of these have been given special names while others have not as shown in Table 2.4. Note that each unit is spelled with a lowercase first letter. If the unit is names for an individual, the first letter of the symbol is capitalized; otherwise, the symbol is lowercase [4].

TABLE 2.4 Derived Dimensions and Their SI Units

Derived Dimension	SI Unit	
	Name	Symbol
Force	newton	N
Energy	joule	J
Power	watt	W
Velocity	-	m/s
Acceleration	-	m/s^2
Density	-	kg/m^3
Pressure or Stress	pascal	Pa
Electric Charge	coulomb	C
Electric Potential	volt	V
Electric Resistance	ohm	Ω
Electric Conductance	siemens	S
Entropy	-	J/K
Inductance	henry	H
Frequency	hertz	Hz

2.9 CONCLUSION

Some of the difficulty that students encounter in thermal technology is a result of the complexity of the systems of units and dimensions. This is particularly true when material properties must be obtained from a variety of sources that use different systems. The objective of this chapter is to acquaint the student with the rationale behind several systems of units and dimensions that have current usage.

Units and dimensions are shown to be abstract quantities which depend on certain natural laws for their definitions. Distinctions between units and dimensions are discussed and the concepts of basic and derived dimensions are defined; symbolism and conversion techniques are also presented. The classification of systems such as the metric vs British and the absolute vs gravitational are described in detail. There are eight different systems covered, and the most commonly used systems are pointed out. Subsequent material and problems throughout the book show how these systems can be applied.

SOLVED PROBLEMS

PROBLEM 1: According to Newton's second law for angular motion, torque equals the product of the mass moment of inertia and angular acceleration ($T = I\alpha$). By means of dimensional symbolism determine the units of I in (a) SI units and (b) British gravitational units.

Solution: $[T] = [FL]$ (i.e., force times length)

$[\alpha] = [T^{-2}]$ (where T stands for time)

$\therefore \quad [I] = [FLT^2]$

(a) SI Units

$[F] = [N], \quad [L] = [m], \quad \text{and} \quad [T] = [s]$

$[I] = [N\ m\ s^2] \quad \text{but} \quad [N] = [kg\ m\ s^{-2}]$

$\therefore \quad [I] = [kg\ m^2]$

(b) British Gravitational Units

$[F] = [lbf], \quad [L] = [ft], \text{ and } \quad [T] = [s]$

$$[I] = [\text{lbf ft s}^2] \quad \text{but} \quad [\text{lbf}] = [\text{slug ft s}^{-2}]$$

$$\therefore \quad [I] = [\text{slug ft}^2]$$

PROBLEM 2: Determine the gravitational constant g_c for the British Absolute system.

Solution: $F = \dfrac{ma}{g_c}$

where

$$[F] = [\text{poundal}]$$

$$[m] = [\text{lbm}]$$

$$[g] = [\text{ft s}^{-2}]$$

One poundal of force is required to accelerate one pound of mass at the rate of one meter per second squared. Therefore, all the elements in the above equation have a unity coefficient:

$$\therefore \quad g_c = 1.0 \text{ lbm ft/poundal s}^2$$

PROBLEM 3: If a mercury barometer shows a height of 76.0 cmHg, express the atmospheric pressure in force units of the British Absolute system.

Solution: $p_a = \rho g h$

where

$$p_a = \text{atmospheric pressure}$$

$$\rho_{Hg} = 847 \text{ lbm/ft}^3$$

$$g = 32.2 \text{ ft/s}^2$$

$$h = 76.0 \text{ cmHg}$$

$$p_a = \left| 847 \frac{\text{lbm}}{\text{ft}^3} \right| \left| 32.2 \frac{\text{ft}}{\text{s}^2} \right| \left| 76.0 \text{ cm} \right| \left| 0.033 \frac{\text{ft}}{\text{cm}} \right| = 68\ 400 \text{ lbm/ft s}^2$$

Mass units can be converted to force units by dividing through by the gravitational constant g_c:

$$p_a = 684\ 000 \frac{\text{lbm}}{\text{ft s}^2} \div 1.0 \frac{\text{lbm ft}}{\text{poundal s}^2} = 68\ 400 \text{ poundal/ft}^2$$

EXERCISES

1. By the use of dimensional symbolism, determine the dimensions of acceleration in the (a) SI system and (b) British gravitational system.
2. What is the gravitational constant in the SI system?
3. Units for surface tension in the cgs system are dynes/centimeter. What are the units in the (a) SI system, (b) British gravitational system, and (c) British Engineering system?
4. Express the following quantities in scientific notation: (a) three and a half million pounds, (b) two trillionths of a second, (c) one hundred gigahertz, and (d) sixteen megapascals.
5. If a mercury barometer reads 76.0 cmHg, express the atmospheric pressure in force units of the (a) SI system and (b) British gravitational system.

REFERENCES

1. H. S. Hvistendahl, *Engineering Units and Physical Quantities*, Macmillan, London, 1964.
2. International Organization for Standardization: Part I, *The International System of Units and Quantities and Units of Space and Time*, ISO/R31, 1958E; Part II, *Quantities and Units of Periodic and Related Phenomena*, ISO/R31, 1958E; Part III, *Quantities and Units of Mechanics*, ISO/R31, 1960E; Part IV, *Quantities and Units of Heat*, ISO/R31, 1960E; Part XI, *Mathematical Signs and Symbols for Use in Science and Technology*, ISO/R31; 1961E.
3. *Letter Symbols for Units Used in Science and Technology*, American Society of Mechanical Engineers, ANSI Y10.19-1969.
4. *Orientation and Guide for Use of SI Units*, ASME Guide SI-I, American Society of Mechanical Engineers, 1975.

BIBLIOGRAPHY

Announcement of Changes in Electrical and Photometric Units, National Bureau of Standards Circular C459, May 15, 1947.

Bridgman, P. W., *Dimensional Analysis*, Yale University Press, New Haven, Conn., 1963

Duncan, W. J., *Physical Similarity and Dimensional Analysis*, Edward Arnolds, London, 1953.

Giorgi, G., *Rational Electromagnetic Units*, AEI 1902 (6), 453.

Handbook of Chemistry and Physics, pp. F-214 to F-236, Chemical Rubber Co., Ohio, 1975.

Kline, S. J., *Similitude and Approximation Theory*, McGraw-Hill, New York, 1965.

Machinery's Handbook, pp. 2400-2440, Industrial Press Inc., 1975.

Rohsenow, W. M. and J. P. Hartnett, *Handbook of Heat Transfer*, pp. 1-15 to 1-17 and 2-3 to 2-7, McGraw-Hill, New York, 1973.

Standard Handbook for Mechanical Engineers, pp. 1-70 to 1-85, McGraw-Hill, New York, 1967.

U. S. Metric Study, Engineering Standards, pp. 168-186, U.S. Dept. of Commerce, National Bureau of Standards, NBS SP 345-11, 1971.

3

TEMPERATURE AND ITS MEASUREMENT

3.1 INTRODUCTION

When heat is applied to a system, the temperature of the system increases; conversely, when heat is extracted, the temperature decreases. Therefore, temperature can be thought of as the property which indicates how much thermal energy is in a system.

A crude way of measuring temperature is to touch an object; the temperature is said to be high when the object feels hot to the touch, and low when it feels cold. When one uses the sense of touch, the amount of heat in an object is not being measured, but rather, the extent of heat transfer from or to the object. This fact formed the basis of an experiment described in 1690 by John Locke, the English philosopher. If three containers of water are prepared--one hot, one cold, and one at an intermediate level--and then if a person places one hand in the hot water and the other in the cold water, and then places both hands in the water of intermediate temperature, the water will seem cold to the first hand and hot to the second hand. Locke concluded from this experiment that heat in a body is characterized by the motion of minute particles and that, under certain conditions, there can be an increase or diminution of this motion. This was a remarkably astute observation since it was made a century and a half before the general scientific community adopted the same viewpoint.

Locke's experiment indicates that an object will be sensed as hot when heat is transferred to the hand and cold when heat is transferred away from it. The human body is, therefore, a reference level above which temperature can be classified as hot and below which it can be classified as cold. The sense of touch can then be used as a thermometer, although an unreliable one.

The unreliability of human sensation as a thermometer is shown by Weber's law, a psychophysical relationship which has been established empirically. The law postulates that sensation is approximately an exponential function of the stimulus that causes it. Figure 3.1 shows that if stimuli increase in the form of a geometric series starting from some threshold value, then the corresponding sensations will increase in the form of an arithmetic series.

Another experiment that has been mentioned in the literature for centuries is the following. If hotter and hotter bodies are touched sequentially, the sensation of pain increases and soon becomes intensely painful. When the pain is intense, an increase in "hottness" of the body has little or no effect on the sensation of pain. Similarly,

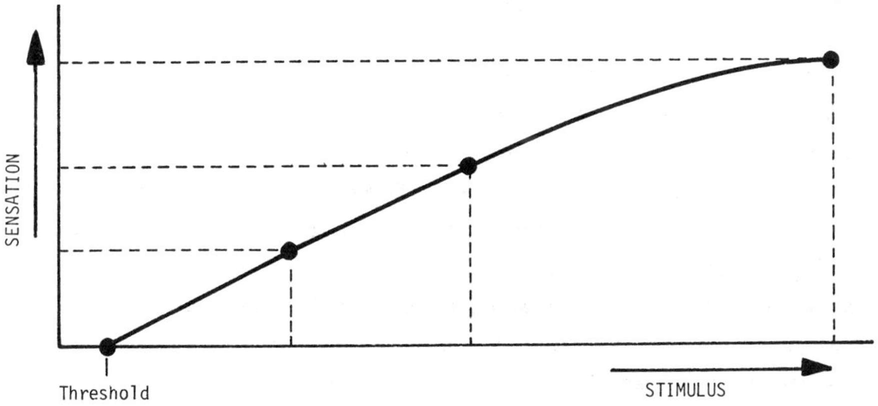

FIGURE 3.1 Weber's law: A psychophysical relationship which states that a physical sensation is approximately an exponential function of the stimulus which causes it. Starting from some threshold value, stimuli increase in the form of a geometric series while sensations increase in the form of an arithmetic series.

if colder and colder bodies are touched sequentially, the sensation of cold increases and becomes intensely painful. An extremely cold body will produce the same sensation of pain as a hot body.

Everyone senses and reacts to heat and heat transfer, although some people's interpretation of thermal phenomena is not always scientific. Consider, for example, the mother who tells her child to wear more outer clothing as a protection against the elements because "your shirt has been washed so often, there is no heat left in it." What she should say, of course, is that the thermal resistance of the shirt has been diminished so that the rate of heat transfer through the shirt is greater than when it was new. Although the mother's interpretation is not precise, her recommendation is sound.

The temperature sense of humans, even when they have a technical knowledge of heat, is unreliable and can even be contradictory. Careful definitions and standards must be established, therefore, for measurements of heat and heat transfer to be used in thermal analyses.

3.2 EQUALITY OF TEMPERATURE

Although temperature is a very basic property, it is not easy to define. One can say that temperature is the property which determines when a system is in thermal equilibrium with a reference. But then, thermal equilibrium is defined in terms of temperature, and a circuitous definition results. It is more convenient to define equality of temperature rather than temperature itself.

If two objects which are at different levels of temperature are brought into intimate contact, certain changes in each of the objects can be observed. For example, changes will occur in properties such as volume, electric resistance, thermal resistance, magnetic state, color, specific heat, density, viscosity, and thermal conductivity. The extent of these changes will depend on the magnitudes of the temperatures. If, however, these objects remain in contact long enough, a condition will be reached where there is no further change in any observable property. When this condition exists, it can be postulated that the two bodies have *equality of temperature*.

3.3 ZEROTH LAW OF THERMODYNAMICS

In addition to equality of temperature, there is another basic principle which is required for the development of practical methods of thermometry. The principle is the *zeroth law of thermodynamics* and states that if two bodies are equal in temperature with a third body, they are equal in temperature with each other. The principle is simple and easily understandable, but it is not trivial or obvious.

3.4 THERMOMETRY

The zeroth law is the basis of temperature measurement. It permits one to use a device called a thermometer, and if a body is in thermal equilibrium with the thermometer, it is considered to be in thermal equilibrium with some specified standard. To be useful, the thermometer must have a consistent and logical scale. Such a scale can be conceived by choosing two standard temperature levels that can be readily reproduced, and then dividing the interval between these fixed points into an appropriate number of equal parts; each division is then called a degree (°).

Thermometers depend upon the change of a physical property as a function of temperature. There are a number of properties which change with temperature consistently enough to be used in practical thermometry:

Volume of a liquid
Electrical resistance of a wire
Length of a rod
Pressure of a gas at constant volume
Volume of a gas at constant pressure
Color of a filament
Electric potential of junctions of dissimilar metals
Magnetic state
Sonic velocity in a gas
Color of thin oxide films
Melting points of paints and crayons

There are five important characteristics to be considered when a thermometer is being selected. *Accuracy* is an indication of how close the temperature reading is to the actual temperature being measured. *Precision* refers to the extent of difference in successive readings of the same temperature. *Sensitivity* is the ratio of the response of the thermometer to some unit change in temperature. *Resolution* relates to the smallest change in temperature to which the thermometer will respond. *Response* is a measure of how long it takes for the thermometer to register a change in temperature.

In addition to the operational characteristics listed above, other important considerations are ruggedness, portability, and cost. No single type of thermometer can perform well under all possible demands, and so it is necessary that the most important demands be identified for any particular application. For example, if a thermometer is to be used as a standard to calibrate other instruments, or if it is to be used to obtain basic research data, the most important characteristics are accuracy, precision, and sensitivity. These characteristics are best satisfied by the gas thermometer.

3.5 DEVELOPMENT OF A TEMPERATURE SCALE

Solids, liquids, and gases will remain in phase equilibrium at only one temperature and pressure. The temperature at which a solid and liquid will coexist under a pressure of one atmosphere is called the normal melting point and is abbreviated NMP; the temperature for the coexistence of liquid and vapor is called the normal boiling point, NBP. Phase equilibrium between solid and vapor will sometimes exist at atmospheric pressure and is called the normal sublimation point, NSP.

The triple point, TP, is the condition in which solid, liquid, and vapor can coexist. The international standard fixed point, established in 1954, was the triple point of water. This point occurs under a pressure of 4.58 mmHg at a temperature of 273.16 K. The symbol K is also an international standard, read as degrees Kelvin. The triple point of water was selected as the international standard because it is the most readily reproducible of all the possible fixed points.

In addition to a fixed point, the development of a temperature scale requires that some means be established for measuring values in relation to the fixed point. This can be done by using any of the thermometric properties listed in the previous section. If we let X stand for any one of these properties, and arbitrarily choose a linear relation between the property and the temperature of a thermometer, then

$$T(X) = aX$$

where a is a constant which must be evaluated. The selection of a linear relationship establishes a condition in which equal temperature intervals will correspond to equal changes in the thermometric property. It should be clear that other functional forms could have been selected if desired.

Any two temperatures will form the same ratio as their corresponding values of X:

$$\frac{T(X)_1}{T(X)_2} = \frac{X_1}{X_2}$$

One of the temperatures could be taken as the triple point of water, i.e., T(X) = 273.16 K, and the corresponding property X_{TP} would be some characteristic property of a thermometer that occurs at the triple point temperature:

$$\frac{T(X)}{T(X)_{TP}} = \frac{X}{X_{TP}}$$

or

$$T(X) = 273.16\ \text{K}\ \frac{X}{X_{TP}} \tag{3.1}$$

In order to have universal agreement on the measurement of temperature, a standard thermometer must be agreed upon. It was stated above that the gas thermometer had the smallest variations in readings, and so this is the recognized standard. In particular, a constant volume gas thermometer is used and the thermometric property is the gas pressure at constant volume.

If the volume of a gas is kept constant, the pressure has a linear dependence on the temperature (a special case is the Gay-Lussac

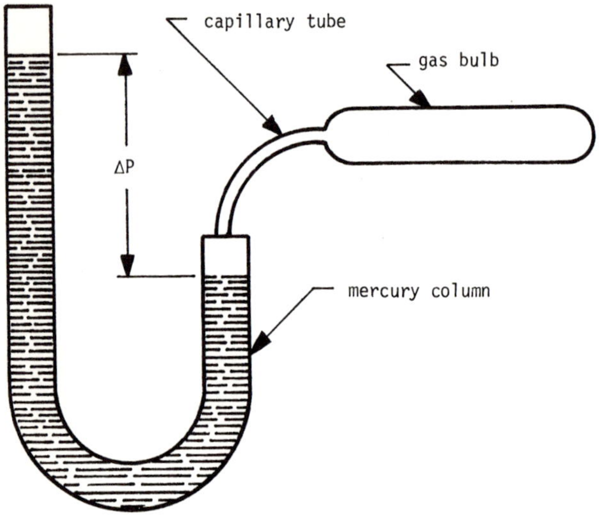

FIGURE 3.2 Schematic illustration of a constant volume gas thermometer.

law for an ideal gas). In a gas thermometer (Fig. 3.2) the gas is contained in a bulb and the pressure exerted by the gas is measured by an open tube manometer which is connected to a mercury reservoir with a flexible tube. The gas bulb is placed in thermal communication with the environment whose temperature is to be determined, and when thermal equilibrium is established, the gas is at the same temperature level as the environment. The volume of the gas will tend to increase or decrease according to whether the gas temperature has been raised or lowered from its initial value; but the volume can be kept constant if the mercury reservoir is raised or lowered. The net result will be a change in the pressure reading which can be related to the triple point temperature by the relation:

$$T(P) = 273.16\ \mathrm{K}\ \frac{P}{P_{TP}} \tag{3.2}$$

In this relation P_{TP} is the pressure exerted by the gas when it is at a temperature of 273.16 K, and P is the pressure measured at the same volume but at a temperature T(P).

3.5 Development of a Temperature Scale

Other thermometric properties can be used in a simila Consider, for example, the ordinary mercury-in-glass thermometer. Mercury is placed in a glass capillary tube and the volume of the mercury changes with temperature. Because of the small area of the bore of the tube, it is more convenient to measure the change in length of the mercury column. The temperature at some particular length T(L) can be expressed in terms of the length of the mercury column when it is at the triple point temperature of water:

$$T(L) = 273.16\ \text{K}\ \frac{L}{L_{TP}} \tag{3.3}$$

Suppose a mercury column is 6.0 cm high when the thermometer is measuring the temperature of the triple point of water. What temperature is being measured when the column is 6.6 cm high?

$$T(L) = 273.16\ \text{K}\ \frac{6.6\ \text{cm}}{6.0\ \text{cm}} = 300\ \text{K}$$

If one particular temperature is measured with several different types of thermometers, the readings obtained will be significantly different. The only temperature that will be the same will be that representing the triple point of water. Table 3.1 shows the results

TABLE 3.1 Comparative Thermometer Readings

FIXED POINT	Copper-Nickel Thermocouple		Platinum Resistance Thermometer		Low Pressure Constant Volume Gas Thermometer (Hydrogen)	
	ε (millivolts)	$T(\varepsilon)$ (K)	R (ohms)	T(R) (K)	P (atmospheres)	T(P) (K)
O_2(NBP)	0	0	2.50	69.5	0.33	90
CO_2(NSP)	1.52	139	6.65	185	0.72	196
H_2O(TP)	ε_{tp} = 2.98	273.16	R_{tp} = 9.83	273.16	P_{tp} = 1.00	273.16
H_2O(NBP)	5.30	486	13.65	380	1.37	374
Sn(NMP)	9.02	826	18.56	516	1.85	505

of a series of measurements of five fixed points by three different thermometers. The thermometers are a copper-nickel thermocouple, a platinum resistance thermometer, and a low-pressure hydrogen gas thermometer. Obviously, to be useful in practical measurements, thermometers must be calibrated to some standard upon which there is universal agreement. The standard thermometer is the constant volume gas thermometer calibrated to an ideal gas temperature scale.

3.6 IDEAL GAS TEMPERATURE

An ideal gas temperature can be defined in the following manner. Let the gas bulb of a constant volume gas thermometer be immersed in water at the triple point and record the gas pressure as $P_{(TP)1}$; next immerse the bulb in condensing steam at 1 atm and record the corresponding gas pressure as P_1. For a second test remove some of the gas from the thermometer bulb so that the triple point pressure reading is $P_{(TP)2}$ and the steam pressure reading is P_2. This procedure is repeated several times while the quantity of gas in the bulb is reduced at each step. At each step the temperature is calculated from the relation

$$T(P) = 273.16\ \text{K}\ \frac{P}{P_{TP}} \tag{3.3}$$

and is plotted versus P_{TP} as shown in Fig. 3.3. If a number of gases are tested, the results plotted, and the curves extrapolated to the axis where $P_{TP} = 0$, it will be found that the temperature readings of all the thermometers will approach the same value of 373.15 K, which is the NBP of water. The extrapolated value of temperature, therefore, depends only on the general properties of gases and not on any particular gas and is defined as the *ideal gas temperature:*

$$T_{\text{ideal}} \equiv 273.16\ \text{K} \lim_{P_{TP}\to 0} \left[\frac{P}{P_{TP}}\right]_{\text{constant volume}} \tag{3.4}$$

3.7 ABSOLUTE THERMODYNAMIC TEMPERATURE SCALE

One objection to the ideal gas temperature is that it is based on properties (namely, the pressure properties of an ideal gas). Since

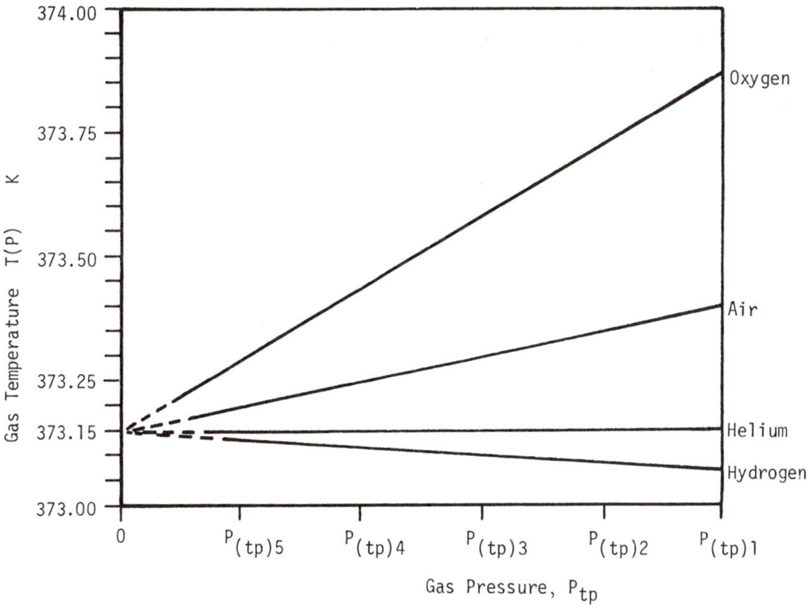

FIGURE 3.3 Absolute temperature as a function of pressure, extrapolated to zero pressure for several gases.

properties are not well behaved at very low temperatures (e.g., gases liquefy), there is some uncertainty about the definition of absolute zero. From a practical engineering viewpoint this is not a matter of great concern, but it is important in the realm of research.

Lord Kelvin proposed a truly absolute thermodynamic temperature scale which is independent of all material properties. The scale is based on the operation of a reversible engine. A Carnot engine operating between two temperature levels is independent of the working substance and depends only on the temperature levels. It will be shown later that the thermal efficiency of a Carnot engine is given by the relation

$$\eta = 1 + \frac{Q_c}{Q_h} \tag{3.5}$$

where Q_c is the heat exchanged with the cold reservoir, and Q_h is the heat exchanged with the hot reservoir. Kelvin's proposal was that the ratio of the reservoir temperatures be arbitrarily set equal to the ratio of the magnitudes of the heat absorbed and rejected:

$$\frac{T_h}{T_c} \equiv \frac{|Q_h|}{|Q_c|}$$

Consider a Carnot engine operating between reservoirs at temperatures T and T_{TP}:

$$\therefore \quad \frac{T}{T_{TP}} = \frac{|Q|}{|Q_{TP}|}$$

or

$$T = 273.16 \text{ K} \frac{|Q|}{|Q_{TP}|} \tag{3.6}$$

It can be seen that the heat is cast in the role of a thermometric property. (It must be kept in mind, however, that heat is not a property.) The value of Q can be made as small as desired, down to a value of zero. When Q is zero the temperature T can be considered as absolute zero.

3.8 CELSIUS AND KELVIN TEMPERATURE SCALES

At the Tenth General Conference on Weights and Measures held in 1954, the Celsius scale was redefined in terms of the triple point of water as a fixed point and was based on the ideal gas temperature scale. As a consequence, the Celsius degree has the same magnitude as the Kelvin (or ideal gas) degree. The triple point of water is assigned the value of 0.01°C, and a secondary standard fixed point, the ice point, is assigned the value 0.00°C. The relation between Celsius and Kelvin scales then becomes:

$$\text{Celsius degrees} = \text{Kelvin degrees} - 273.15 \tag{3.7}$$

The steam point can be found from the experimental data plotted in Figure 3.3:

$$\text{Celsius degrees at the steam point} = 373.15 \text{ K} - 273.15 = 100.00°\text{C}$$

3.9 RANKINE AND FAHRENHEIT TEMPERATURE SCALES

Gabriel Fahrenheit (1686-1736), a German instrument maker who did most of his work in Holland, is generally credited with the invention of the mercury-in-glass thermometer and the development of the temperature scale that bears his name. Fahrenheit originally chose two fixed points: the temperature of a certain mixture of salt and ice to which he assigned a value of zero, and the temperature of the human body which he set at 96°. On this scale the ice point was approximately 32°. The scale was later revised slightly so that the ice point became exactly 32 and the steam point was 212. (On the revised scale, the average body temperature was 98.6).

The interval between the steam point and ice point is 180 on the Fahrenheit scale and 100 on the Celsius scale. Clearly then, one Fahrenheit degree is 5/9 as large as a Celsius degree and the relationship between the two can be established:

$$\text{Celsius degrees} = \frac{5}{9}\,(\text{Fahrenheit degrees} - 32) \tag{3.8}$$

$$\text{Fahrenheit degrees} = \left(\frac{9}{5}\,\text{Celsius degrees}\right) + 32 \tag{3.9}$$

Absolute temperature corresponding to the Fahrenheit scale is measured on the Rankine scale. The relationship between the two scales is given by the following equation:

$$\text{Fahrenheit degrees} = \text{Rankine degrees} - 459.67 \tag{3.10}$$

A comparison of the four most commonly used temperature scales is shown in Fig. 3.4, and Table 3.2 lists the fixed point temperatures that have received international agreement.

3.10 THERMAL EXPANSION

The most common effects of temperature changes are changes in size and changes of state of materials.

Each of the atoms or molecules in a given amount of matter can be visualized as connected to each of its neighbors by means of a

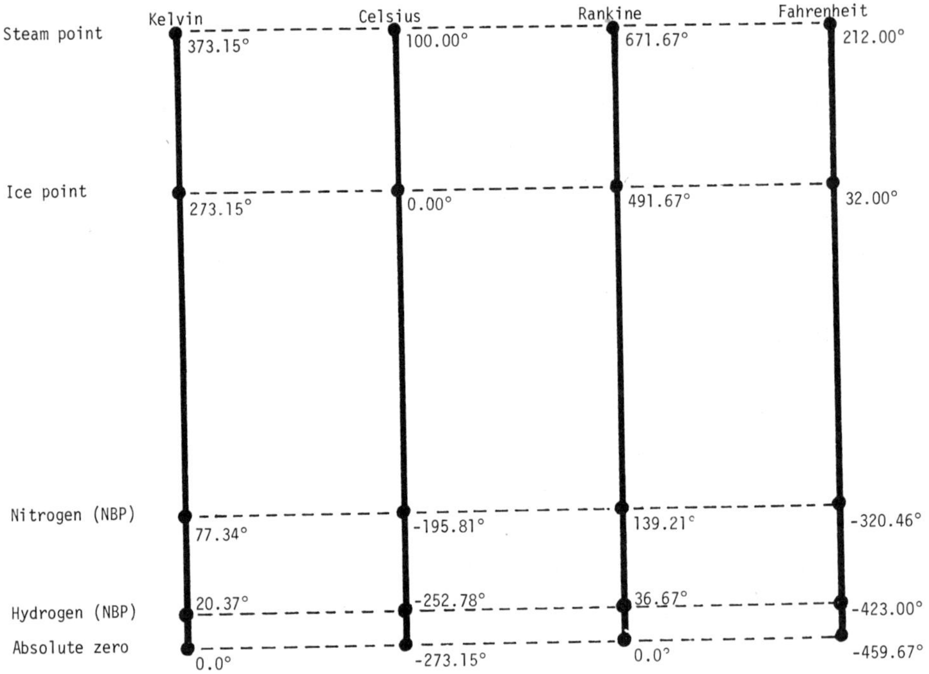

FIGURE 3.4 Comparison of Kelvin, Celsius, Rankine, and Fahrenheit scales.

flexible connector. This connector can be extended or contracted by the action of gravitational, pressure, electrical, magnetic, or thermal fields, and the characteristics of the connector can be changed by various kinds of chemical and physical phenomena.

When a thermal field causes an increase in temperature, the amplitude of vibration of the molecules increases. This leads to an increase in size of a particular quantity of matter.

In the case of solids there can be a linear or one-dimensional expansion. This will occur when one dimension of a heated body is very much greater than the other two. If the length is designated as ℓ, the change in length arising from a change in temperature ΔT is $\Delta\ell$. It has been found experimentally that this change in length is proportional to the original length and the temperature change:

$$\Delta\ell = \alpha\ell\ \Delta T \tag{3.11}$$

TABLE 3.2 International Fixed Point Temperatures

	FIXED POINT	kelvin degrees	celsius degrees	rankine degrees	fahrenheit degrees
Standard	Water (TP)	273.16	0.01	491.688	32.018
Basic	Ice point	273.15	0.00	491.67	32.00
	Steam point	373.15	100.00	671.67	212.00
	Sulfur (NBP)	717.75	444.60		
	Antimony (NMP)	903.65	630.50	1626.57	1166.90
	Silver (NMP)	1233.95	968.80	2221.11	1761.44
	Gold (NMP)	1336.15	1063.00	2405.07	1945.40
Secondary	Helium (NBP)	4.22	-268.93		
	Hydroger (NBP)	20.37	-252.78		
	Nitrogen (NBP)	77.34	-195.81		
	Mercury (NMP)	234.29	-38.86		
	Lead (NMP)	600.45	327.30		
	Zinc (NMP)	692.65	419.50		

The proportionality constant α is called the *temperature coefficient of linear expansion*. The value of α is temperature dependent, but for most engineering applications the variation is negligible and constant values can be used in calculations. Linear coefficients of expansion are determined experimentally and some typical values are shown in Table 3.3.

For isotropic solids the change in area due to a temperature change can be found approximately by assuming a coefficient of 2α:

$$\Delta A \simeq 2\alpha A \, \Delta T \tag{3.12}$$

and, correspondingly, for changes in volume:

$$\Delta V \simeq 3\alpha V \, \Delta T \tag{3.13}$$

TABLE 3.3 Average Values of Linear Thermal Coefficient of Expansion around Room Temperature

Material	$\alpha \times 10^6$ / $°C^{-1}$
Aluminum (2024-T3)	22.7
Brass Wire	19.3
Concrete	14.4
Copper (pure)	16.6
Glass (ordinary)	9.0
Glass (pyrex)	3.2
Invar	0.8
Platinum	9.0
Silver (pure)	19.8
Steel (AISI C1020)	15.1
Steel (AISI 304)	17.8

When calculating volume changes, it is customary to use the *volume coefficient of expansion* β, particularly for fluids. These coefficients are also determined experimentally and typical values are shown in Table 3.4:

$$\Delta V = \beta V \ \Delta T \tag{3.14}$$

If there is a hole in a solid body, the volume of the hole increases when the body expands just as if the hole were a solid of the same material as the body. This is true even if the hole becomes so large that the surrounding body is reduced to a thin shell. Thus, the volume enclosed by a thin-walled glass flask or a thermometer bulb increases in the same way as would a solid glass body of the same size.

Suppose, for example, that a pyrex glass flask whose volume of 100 cm^3 is filled with mercury at room temperature. Determine how much mercury will overflow if the temperature is raised 50°C. This

TABLE 3.4 Average Values of Volume Thermal Coefficient of Expansion around Room Temperature

Material	$\frac{\beta \times 10^{-6}}{°C^{-1}}$
Aluminum	72.0
Brass	60.0
Copper	42.0
Glass (pyrex)	9.5
Mercury	180.0
Platinum	27.0
Silver	58.3
Wood (pine)	50.4

problem can be solved by assuming that the volume of the glass flask and the mercury expand according to Eq. (3.14) and that the values of β given in Table 3.4 are correct over the temperature range encountered in the problem.

The increase in volume of the flask is

$$\Delta V = (9.5 \times 10^{-6}\ °C^{-1})(100\ cm^3)(50°C) = 0.048\ cm^3$$

while the increase in volume for the mercury is

$$\Delta V = (180 \times 10^{-6}\ °C^{-1})(100\ cm^3)(50°C) = 0.9\ cm^3$$

Therefore, the overflow is the difference between the two volume increases:

$$\text{Overflow} = 0.9 - 0.048 = 0.85\ cm^3$$

A number of special alloys have been made which have expansion coefficients of certain desired values. Invar (64 w/o Fe 36 w/o Ni) has a linear coefficient of expansion of only $0.8 \times 10^{-6}\ °C^{-1}$. It is often used with copper to form elements in thermostats. Platinite (54 w/o Fe 46 w/o Ni) has the same coefficient as platinum and is sometimes used as a substitute for that metal in electrical work.

Also, it has practically the same coefficient as the glass used in vacuum tubes and, thus, can be used as a lead wire.

The shape of a fluid is not definite; only volume changes can be considered when temperatures are raised or lowered. According to Charles' Law (defined in Chap. 8) the relative increase in volume for most gases is the same for each degree increase in temperature.

Practically all liquids expand with increasing temperature. A notable exception is water, which decreases in volume when the temperature is increased in the range from 0 to 4°C. In this range the coefficient of expansion is negative. The maximum density of water occurs at 4°C and this is the reference point for the measurement of specific gravity (see Table 3.5).

Both linear and volume coefficients of expansion are highly dependent on temperature. Therefore, it should be noted that Tables 3.3 and 3.4 list average values of the coefficients at about room temperature.

TABLE 3.5 Density of Water as a Function of Temperature

t(°C)	ρ(g/ml)
0	0.999868
1	0.999927
2	0.999968
3	0.999992
*4	1.000000
5	0.999992
6	0.999968
7	0.999930
8	0.999877

*The temperature of maximum density of pure water, free from air, actually occurs at about 3.98°C. At that temperature, the density is 0.999 973 g/cm^3, or 1.000 000 g/ml, based on 1 liter = 1000.028 cm^3.

3.11 CONCLUSION

Since temperature is probably the most important property in thermal technology, this chapter acquaints the student with the basic concepts and standards for temperature and its measurement.

Equality of temperature and the Zeroth law of thermodynamics are discussed, and the elements of thermometry are introduced. The combined use of thermometric properties and mathematical relationships in the development of temperature scales is described. Several temperature scales used in thermal technology are outlined.

Thermal expansions in one, two, and three dimensions are described, along with analytical methods for predicting their magnitudes.

SOLVED PROBLEMS

PROBLEM 1: Make the following conversions: (a) -3°C to °F, (b) 650°R to °C, and (c) 650°R to K.

Solution:

(a) $°F = \frac{9}{5}°C + 32 = \frac{9}{5}(-3°C) + 32 = 26.6°F$

(b) $°F = °R - 459.67 = 650 - 459.67 = 190.33°F$

$$°C = \frac{5}{9}(°F - 32) = \frac{5}{9}(190.33 - 32) = 87.96°C$$

(c) $°C = K - 273.15$

$$K = 87.96 + 273.15 = 361.11\ K$$

PROBLEM 2: The temperature of an object drops from 40°C to 30°C. Determine the temperature drop in (a) °F and (b) °R.

Solution:

(a) 40° - 30° = 10°C, but one Celsius degree is 9/5 (or 1.8) as large as one Fahrenheit degree,

$$\therefore\ (10°C)(1.8) = 18°F \text{ temperature drop}$$

(b) Since °C + constant = K, a temperature difference in °C is the same as a difference in K. Also, one Kelvin degree is 1.8 times as large as one Rankine degree,

$$\therefore \quad 40° - 30° = 10°C = 10\ K$$

$$(10\ K)(1.8) = 18°R \text{ temperature drop}$$

PROBLEM 3: A brass wire has a length of 400 mm at 20°C. What is the length of the wire at 90°C?

Solution: Use data from Table 3.3 and assume that α remains constant between 20 and 90°C:

$$\Delta \ell = \alpha \ell \ \Delta t$$

where

$\Delta \ell$ = change in length

$\alpha = 19.3 \times 10^{-6}\ °C^{-1}$

$\ell = 400$ mm

$\Delta t = 90 - 20 = 70°C$

$$\Delta \ell = (19.3 \times 10^{-6})(400\ \text{mm})(70) = 0.54\ \text{mm}$$

$$\therefore \quad \text{the length at } 90°C = 400.54\ \text{mm}$$

PROBLEM 4: An Al 2024-T3 plate has a 2.980-in diameter hole at 69°F. What is the diameter if the temperature is lowered to -15°F?

Solution: Use data from Table 3.3 and assume that α remains constant between 68° and -15°F:

$$\text{initial area of hole } A_0 = \frac{\pi(2.980)^2}{4} = 6.97\ \text{in}^2$$

$$\alpha \text{ for Al 2024-T3} = 22.7 \times 10^{-6}\ °C^{-1}$$

$$\frac{22.7 \times 10^{-6}\ °C^{-1}}{1.8} = 12.6 \times 10^{-6}\ °F^{-1}$$

Assume that the material is isotropic so that the change in area can be found by using 2α:

$$A \simeq 2\alpha A_0\ \Delta t = 2(12.6 \times 10^{-6})(6.97)(68 - (-15)) = 0.0146\ \text{in}^2$$

$$\text{area of hole at } -15°F = 6.97 - 0.0146 = 6.955\ \text{in}^2$$

$$\therefore \quad \text{diameter at } -15 = \left(\frac{(4)(6.955)}{\pi}\right)^{1/2} = 2.975\ \text{in}$$

EXERCISES

1. What type of thermometer is used for the international standard?
2. What is an objection to the use of a gas thermometer?
3. What thermometric property is used in the absolute thermodynamic temperature scale?
4. Make the following conversions: (a) -3°F to °C, (b) 250°R to °C, and (c) 250 K to °R.
5. The atmospheric temperature has changed from 40°F to -18°F; determine the temperature drop in (a) °C and (b) K.
6. An aluminum wire has a length of 3 ft at 70°F; determine the length at (a) 150°F and (b) -50°F. What assumptions must be made?
7. What is the change in area of a 10-in^2 hole in a copper plate if the temperature changes from 70°F to 200°F?
8. Draw a sketch of a bimetallic strip of copper and Invar after the temperature is raised from 68°F to 150°F.
9. An aluminum sphere has a diameter of 1 in at 68°F. If the temperature is raised to 190°F, determine (a) the surface area and (b) the volume.

BIBLIOGRAPHY

Benedict, R. P., *Temperature and its Measurement*, Electro-Technology, C-M Technical Publications, 71 pp., July 1963.

Burgess, G. K., The international temperature scale, *J. Res. NBS*, RP 22, 635 pp., July-Dec. 1928.

Joule, J. P. and W. Thomson, *On the Thermal Effects of Fluids in Motion*, Phil. Trans. Roy. Soc. London, Vol. 144, 350 pp., June 15, 1854.

Mott-Smith, M., *Heat and its Workings*, Dover Publications, New York, 1962.

Stimson, H. F., The international temperature scale of 1948, *J. Res. NBS*, RP 1926, 209 pp., March 1949.

Thomson., *On an Absolute Thermometric Scale Founded on Carnot's Theory of the Motive Power of Heat*, Cambridge Phil. Soc. Proc., June 5, 1848.

4

VISCOSITY

4.1 INTRODUCTION

Viscosity is the property of a fluid that is a measure of its resistance to flow. All real fluids are viscous; consequently, frictional phenomena are exhibited in all fluid flow. As one layer of a flowing fluid moves past an adjacent layer, there is an interaction of the molecules at the interface due to cohesion. This interaction causes a transport of momentum, which must be conserved and which tends toward equilibrium. Therefore, viscosity can also be considered as the tendency for two contiguous layers of flowing fluid to assume the same velocity.

4.2 COEFFICIENT OF VISCOSITY

In developing the fundamental equation for viscous flow, it is convenient to consider one type of apparatus which is used for measuring the viscosity of a liquid. The apparatus is illustrated in Fig. 4.1 and consists of a cylinder which can rotate coaxially within a cylindrical vessel. The test liquid is placed in the annular space between the cylinders. When the weight is released the inner cylinder will accelerate initially and then continue to rotate at a constant angular velocity when equilibrium is established. The terminal velocity will

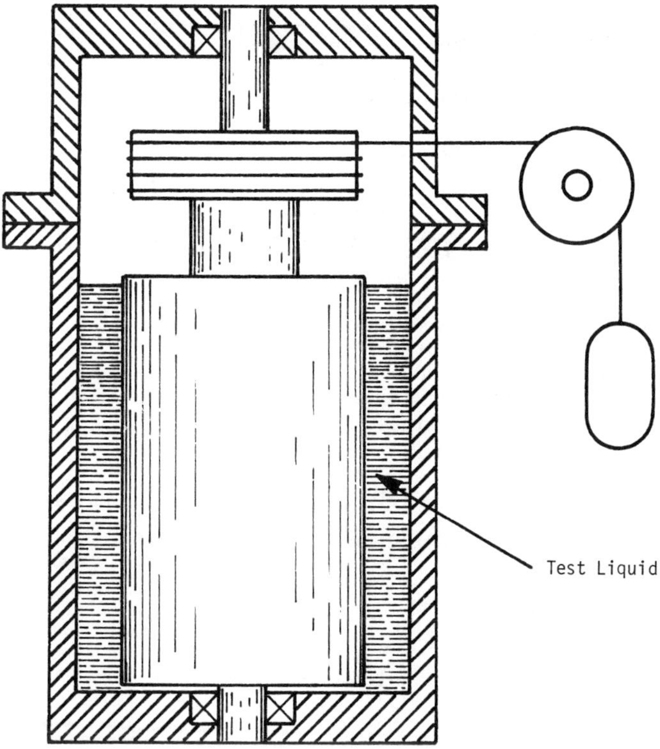

FIGURE 4.1 Schematic drawing of a viscometer.

be a function of the resistance the liquid offers to the rotation of the inner cylinder.

If the cylinders in the apparatus are close to each other in size, the annulus will be thin and a short arc can be represented as a straight section as shown in Fig. 4.2. The inner cylinder and the layer of liquid adjacent to it move at a uniform velocity u, while the outer cylinder and its liquid layer are at rest. Equal and oppositely directed forces F act to maintain equilibrium. If the area over which these forces act is A, the shear stress exerted on the liquid can be expressed as the ratio $F/A = \tau$. For relatively small values of y and u, the velocity gradient between the cylinders will be linear, where v is the velocity at any point.

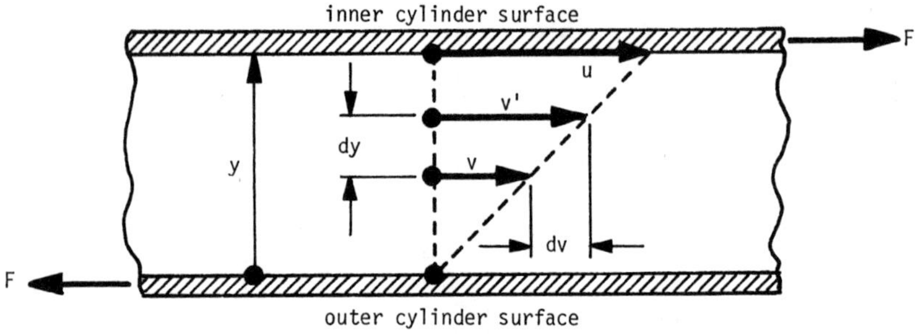

FIGURE 4.2 Laminar flow of a viscous liquid in the annular space between the cylinders of Figure 4.1. When dimension y is small compared with the cylinder diameters, the annular section can be assumed to be bound by two parallel, straight lines.

It has been shown experimentally that the force F will vary directly as the area A and the velocity u, and inversely as the annular dimension, i.e.,

$$F \propto \frac{Au}{y}$$

but u/y = dv/dy (from similar triangles) and, therefore,

$$\frac{F}{A} = \tau \propto \frac{dv}{dy}$$

The proportionality relation can be changed to an equality relation if a proportionality constant μ is placed in the equation:

$$\tau = \mu \frac{dv}{dy} \tag{4.1}$$

In words, this equation states that the shearing stress is proportional to the velocity gradient or the rate of deformation. The constant μ is the *absolute* (or dynamic) coefficient of viscosity and is a property of the fluid. The type of motion illustrated in Fig. 4.2 is called *laminar flow* and is characterized by one layer of liquid sliding over another in the manner that a deck of cards might deform when subjected to a shearing stress. Equation (4.1) was first postulated by Newton, and so liquids which flow in this manner are called *Newtonian fluids*. There are other types of fluids which do not exhibit the linear proportionality between shear stress and velocity gradient and they are referred to as *Nonnewtonian fluids*.

An alternate viscosity coefficient which is independent of dynamic considerations is the *kinematic viscosity*. This coefficient ν is defined as the ratio of the dynamic viscosity and the density of the fluid:

$$\nu \equiv \frac{\mu}{\rho} \tag{4.2}$$

4.3 UNITS OF VISCOSITY

Units for dynamic and kinematic viscosity coefficients can be derived from Eqs. (4.1) and (4.2). SI units have been assigned but, unfortunately, much of the literature continues to express viscosity in the British and cgs units. Consequently, the three sets of units will be described here.

SI Units for Viscosity

Keeping in mind that the SI unit for stress is the pascal (Pa), which is a newton per square meter (N/m^2), the units for the dynamic viscosity coefficient μ can be found from Eq. (4.1):

$$\mu = \frac{dy\ \tau}{dv} = \frac{[m][Pa]}{[m/s]} = [Pa\ s]$$

The dynamic coefficient can also be expressed in terms of mass if the gravitational constant g_c is applied:

$$\mu g_c = \frac{N\ s}{m^2}\frac{kg\ m}{N\ s^2} = \frac{kg}{m\ s}$$

The kinematic coefficient can be found from Eq. (4.2):

$$\nu = \frac{\mu}{\rho} = \frac{kg}{m\ s}\frac{m^3}{kg} = \frac{m^2}{s}$$

British Units for Viscosity

In a manner similar to that used for SI units, the British units for viscosity can be found from Eqs. (4.1) and (4.2):

$$\mu = \frac{dy\ \tau}{dv} = \frac{[ft][lbf/ft^2]}{[ft/s]} = \left[\frac{lbf\ s}{ft^2}\right]$$

$$\mu g_c = \left[\frac{\text{lbf s}}{\text{ft}^2}\right]\left[\frac{\text{lbm ft}}{\text{lbf s}^2}\right] = \left[\frac{\text{lbm}}{\text{ft s}}\right]$$

$$\nu = \frac{\mu}{\rho} = \left[\frac{\text{lbm}}{\text{ft s}}\right]\left[\frac{\text{ft}^3}{\text{lbm}}\right] = \left[\frac{\text{ft}^2}{\text{s}}\right]$$

Care must be exercised when g_c is applied to British units. Customarily, British engineering units are used in thermal technology so that g_c = 32.2 lbm ft/lbf s^2:

$$1.0 \text{ lbm/ft s} = 32.2 \text{ lbf s/ft}^2$$

Also, in some of the literature a unit called the reyn is used to express dynamic viscosity:

$$1 \text{ reyn} \equiv 1 \text{ lbf s/in}^2$$

cgs Units for Viscosity

cgs units for viscosity can be found in a similar manner:

$$\mu = \frac{dy\ \tau}{dv} = \frac{[\text{cm}][\text{dyn/cm}^2]}{[\text{cm/s}]} = \left[\frac{\text{dyn s}}{\text{cm}^2}\right]$$

$$\mu\ g_c = \left[\frac{\text{dyn s}}{\text{cm}^2}\right]\left[\frac{\text{gm cm}}{\text{dyn s}^2}\right] = \left[\frac{\text{gm}}{\text{cm s}}\right]$$

Both the force and mass forms of the dynamic viscosity coefficient in the cgs system are called a *poise*. This is not a serious inconsistency since the gravitational constant g_c has a magnitude of 1.0 in that system:

$$\nu = \frac{\mu}{\rho} = \left[\frac{\text{gm}}{\text{cm s}}\right]\left[\frac{\text{cm}^3}{\text{gm}}\right] = \left[\frac{\text{cm}^2}{\text{s}}\right]$$

The kinematic viscosity coefficient in the cgs system is called a *stoke*. In practice, the special names in this system are customarily taken as the *centipoise* (poise/100) and the *centistoke* (stoke/100).

It is unfortunate that the units for viscosity present such a complex picture. The reader can establish an orderly approach by understanding the basic difference between absolute (or dynamic) viscosity and kinematic viscosity, and by keeping in mind that the

TABLE 4.1 Basic Units for Viscosity Coefficient in Three Systems of Units

Type of Viscosity	Symbol	SI	British	cgs	
Dynamic (Absolute)	μ	Pa s or kg/m s	lbf s/ft^2 or lbm/ft s	dyne s/cm^2 or g/cm s	poise
Kinematic	ν	m^2/s	ft^2/s	cm^2/s	stoke

Relationship between units

Relationship	
[Pa s] x (10) = [poise]	Dynamic Viscosity
[Pa s] x (2.088 x 10^{-2}) = $[\frac{\text{lbf s}}{\text{ft}^2}]$	
$[\frac{\text{lbf s}}{\text{ft}^2}]$ x (4.788 x 10^2) = [poise]	
$[\frac{\text{lbf s}}{\text{ft}^2}]$ x (32.2) = $[\frac{\text{lbm}}{\text{ft·s}}]$	
$[\frac{\text{m}^2}{\text{s}}]$ x (10^4) = [stoke]	Kinematic Viscosity
$[\frac{\text{m}^2}{\text{s}}]$ x (10.76) = $[\frac{\text{ft}^2}{\text{s}}]$	
$[\frac{\text{ft}^2}{\text{s}}]$ x (929) = [stoke]	

absolute quantity can be expressed in terms of both mass and force. Table 4.1 shows the basic units used in three different systems.

Many compilations of viscosity coefficients for fluids are available in the literature. A few of the more complete ones can be found in Refs. 1-5.

4.4 SAYBOLT VISCOSITY

The determination of dynamic viscosity is difficult and usually limited to laboratory conditions. As a consequence, the American Society for Testing Materials (ASTM) has adopted the Saybolt viscometer for quick determinations of viscosity. The methods described in ASTM D88-56 are empirical measurements of the Saybolt viscosity of petroleum products between the temperatures of 70 and 210°F. The technique consists of a determination of the efflux time in seconds of 60 ml of sample flowing through a calibrated orifice. Two different size orifices are defined to measure Saybolt Universal Seconds (SUS) and Saybolt Furol Seconds (SFS). "Furol" is a contraction of fuel and road oils and designates viscosities greater than 1000 SUS. The Saybolt universal viscometer is used for viscosities equal to or greater than 32 SUS, whereas the Saybolt furol viscometer is not recommended for viscosities less than 25 SFS.

Outside the United States, Engler and Redwood viscometers are used. The equations shown in Table 4.2 can be used to obtain kinematic viscosities from each of the methods mentioned in this section.

4.5 EFFECTS OF TEMPERATURE AND PRESSURE

It was pointed out previously that the resistance to shear stress which results from viscosity is caused by two phenomena: cohesion of

TABLE 4.2 Conversions from Efflux Times of Several Types of Viscometers to Kinematic Viscosity

Engler	$\nu = 0.00147\,t - \frac{3.74}{t}$
Redwood	$\nu = 0.0026\,t - \frac{1.715}{t}$
Saybolt Universal	$\nu = 0.0022\,t - \frac{1.80}{t}$
Saybolt Furol	$\nu = 0.0222\,t - \frac{2.03}{t}$

ν = kinematic viscosity in stokes

t = efflux time in seconds

the molecules and momentum transfer from one layer to another. In the case of liquids, the cohesion phenomenon predominates, and since cohesive forces decrease with increasing temperature, the viscosity also decreases. For gases the momentum transfer is more important, and since this phenomenon increases with increasing temperature, the viscosity of gases also increases with temperature.

Absolute viscosity is independent of pressure in the ranges normally encountered in engineering applications. Kinematic viscosity, however, is inversely proportional to density and, consequently, it is inversely proportional to the pressure. Therefore, when kinematic viscosity is being used, both temperature *and* pressure should be specified.

There is an interesting result of using two types of viscosities, i.e., dynamic and kinematic. A liquid such as water is more viscous than air in terms of dynamic viscosity, but air is more viscous than water in terms of kinematic viscosity.

4.6 NONNEWTONIAN FLUIDS

All gases and most liquids used in engineering applications are Newtonian in nature and thus obey the rules described above. There are, however, certain pastes, slurries, and polymers which do not show the linear relation between shear stress and velocity gradient; these fluids are included in the field of study called *rheology*. In addition to the flow of nonnewtonian liquids, the field of rheology also covers the plastic flow of solids.

Three types of nonnewtonian fluids are classified: plastic, pseudoplastic, and dilatant. A qualitative understanding of the flow associated with each of these types of fluid, and the comparison with Newtonian fluids, can be shown by rheograms. The rheograms of Fig. 4.3 are plots of the rate of shear vs shear stress, and dynamic viscosity vs rate of shear. It is evident from Eq. (4.1) that dynamic viscosity is the reciprocal of the slope of the former rheogram.

In a plastic substance (Fig. 4.3b) a threshold value of shear stress must be reached before there is any shear or flow. As the rate

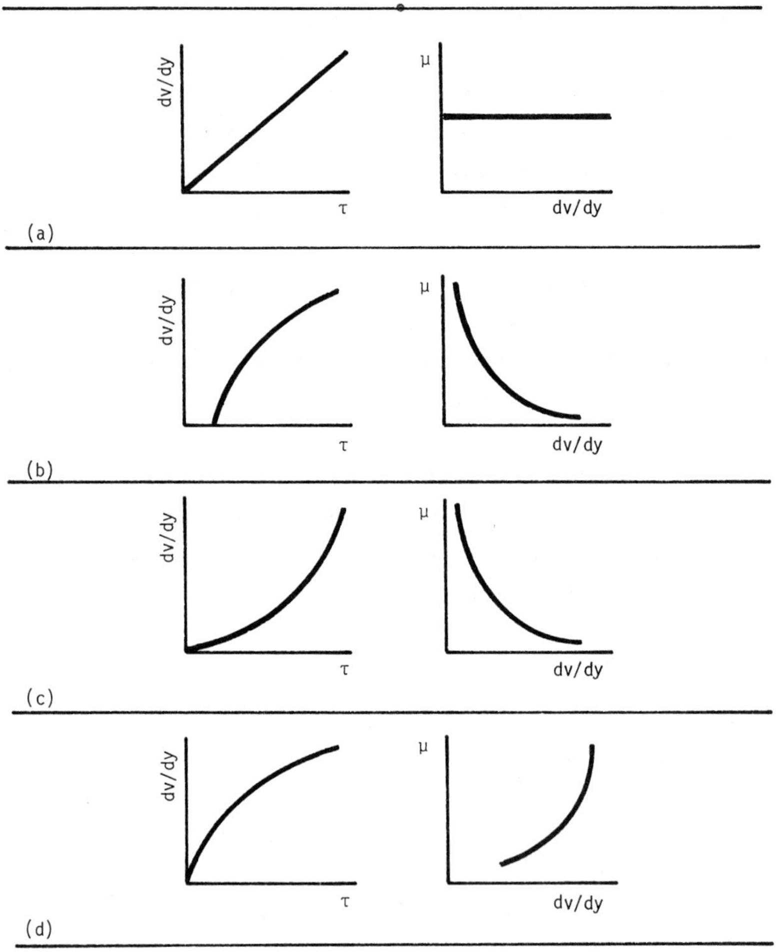

FIGURE 4.3 Rheograms for four kinds of material [rate of shear (dv/dy) vs shear stress (τ) and dynamic viscosity (μ) vs rate of shear (dv/dy)]: (a) Newtonian fluid, (b) plastic material, (c) pseudoplastic material, and (d) dilatant material.

of shear approaches zero, the dynamic viscosity approaches infinity. Fluids with these characteristics are said to be thixotropic. *Thixotropy* is a property which causes substances to be fluid when they are stirred or shaken and to act as solids when they are at rest. A practical example of a thixotropic substance is the type of paint which is very thick when it is in the can, but which can be spread very easily with a brush. Other examples are materials packaged in tubes, such as certain greases, salves, and toothpaste; these materials cannot be poured out of the tube, but they can be easily squeezed from the tube.

In a pseudoplastic substance (Fig. 4.3c) the viscosity coefficient decreases as the rate of shear increases, but there is no threshold value of shear stress. Consequently, pseudoplastic materials can always be considered as fluids. They are examplified by various kinds of suspensions.

A dilatant material (Fig. 4.3d) shows an increase in resistance to deformation as the rate of shear increases. This is the direct opposite of a plastic material. It is possible for the viscosity to reach a high enough value so that the fluid effectively changes to a solid. It is important to know when a fluid is dilatant in nature if a pump is used. As the pump is sped up, the dilatant fluid can conceivably solidify and cause damage to the pump.

4.7 EFFECTS OF VISCOSITY ON FLUID FLOW

Viscosity is one of the more important properties that determine the manner in which a fluid will flow. Inclusion of viscous effects can increase the mathematical complexity of an analysis, and so sometimes an ideal fluid is postulated. An ideal fluid is inviscid, i.e., there are no viscous effects. One should always keep in mind, however, that all real fluids exhibit viscosity.

Boundary Layer Concept

As a fluid flows past a surface, it can be observed that the layer of fluid adjacent to the surface has zero relative velocity. As the motion is traced farther out in the stream, it can be noted that the relative velocity increases in proportion to the distance from the surface. The velocity gradient is a direct effect of the viscosity of the fluid. At some distance from the surface there will be no further change in velocity, and it can be assumed that all viscous forces have subsided. This distance delineates the thickness of the *hydrodynamic boundary layer* δ_h, and is defined as the distance required for the fluid to attain 99% of the free stream velocity. The hydrodynamic boundary layer is illustrated schematically in Fig. 4.4. In Chap. 12 on Convection Heat Transfer an analogous concept will be used to study the thermal properties of a fluid system. Since the relative velocity of fluid particles at the surface is zero, heat is transferred by conduction only. Since, with the exception of liquid metals, the thermal conductivity of fluids is relatively low, there is a large temperature gradient at the wall. Farther out in the layer, fluid motion reduces the gradient until it becomes zero. The distance required for the temperature to attain 99% of the free stream temperature is called the *thermal boundary layer* δ_t and is illustrated in Fig. 4.5.

When discussing boundary layers one must consider two broad classifications of fluid flow: external and internal flow. In *external flow* a surface is immersed in an unbounded fluid; in *internal flow* a fluid is completely bounded by solid surfaces.

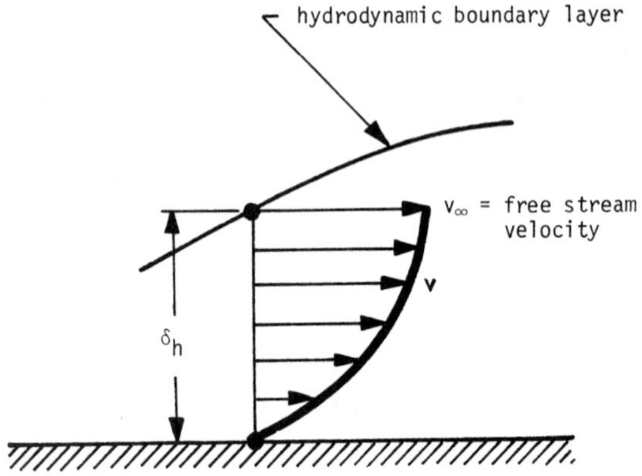

FIGURE 4.4 Fluid velocity profile in a hydrodynamic boundary layer.

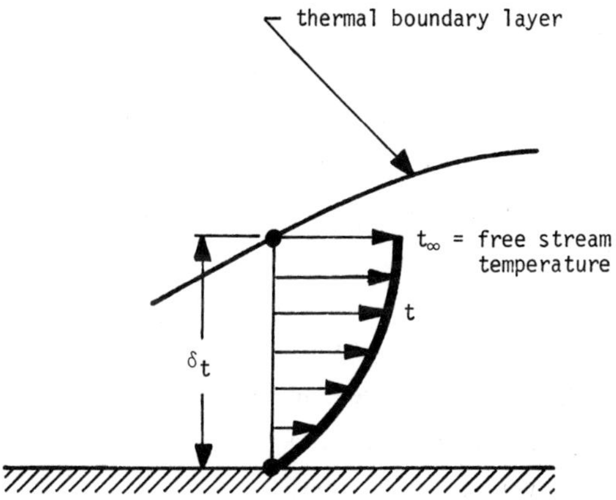

FIGURE 4.5 Fluid temperature profile in a thermal boundary layer.

External Flow: Figure 4.6 shows an external flow of a fluid over a flat plate. The fluid approaches the plate with a velocity v . As the fluid passes over the plate, viscous effects establish a velocity gradient; as the fluid proceeds along the plate, the viscous effects cause the boundary layer to thicken. At some critical distance from the leading edge there is an amplification of small disturbances in the fluid and the nature of the flow changes. This change introduces

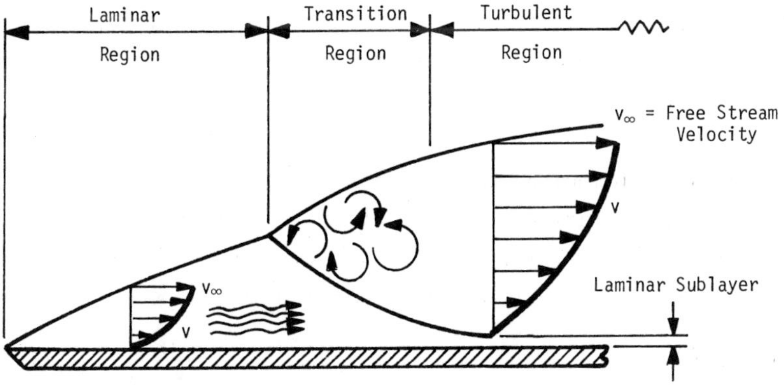

FIGURE 4.6 Boundary layer flow regimes for a fluid in external forced flow over a flat plate (vertical scale exaggerated).

two other broad classifications of fluid flow: laminar and turbulent flow.

In *laminar flow* there is no macroscopic mixing of adjacent fluid layers; in *turbulent flow* there is a macroscopic mixing of fluid particles and velocity fluctuations are superimposed on the general motion of the fluid (see Fig. 4.6).

The change from laminar to turbulent flow does not occur suddenly, but is spread out over an area of the plate that is called the transition region. Even in the turbulent region there is a thin layer of fluid that remains laminar in nature. In the fluid outside the boundary layer, velocity gradients are very small and viscosity effects are negligible.

Internal Flow: Figure 4.7 illustrates the type of boundary layer formed in internal flow. The fluid enters the conduit with a uniform velocity and a boundary layer of low velocity begins to form at the entrance. This layer thickens along the flow path because of viscous effects. A unique and simplifying feature of this type of flow is that the boundary layers converge on the centerline of the tube at some distance from the entrance. This distance is called the *hydrodynamic entry length* and establishes the region in which the fluid velocity is a function of the length along the centerline; it also establishes the beginning of the fully developed flow region. The actual shape of the fully developed velocity profile depends on whether the flow is laminar or turbulent.

Reynolds Number

The transition from laminar to turbulent flow, or the existence of a fully developed velocity profile, is of great interest in the analysis of fluid or thermal systems. Such predictions are possible through the use of a nondimensional Reynolds number Re:

$$\mathrm{Re} \equiv \frac{\rho v D}{\mu} \tag{4.3}$$

In this relation, ρ is the fluid density, v is a characteristic fluid velocity, D is some characteristic dimension, and μ is the

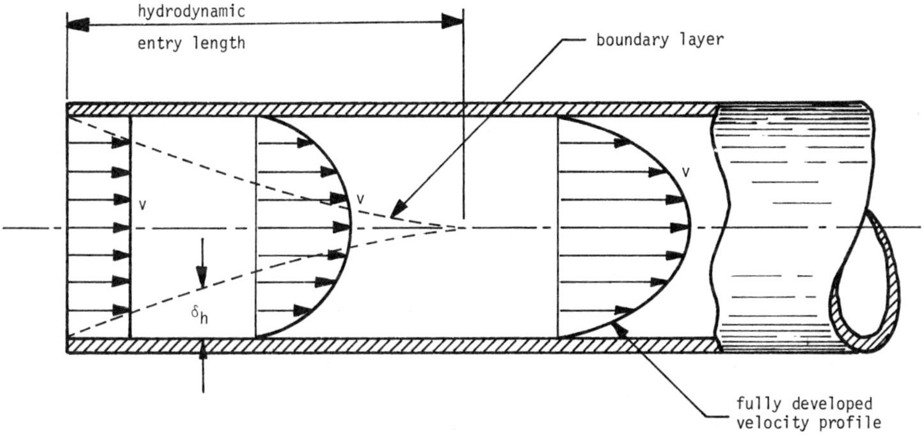

FIGURE 4.7 Boundary layer flow regimes for internal flow of a fluid through a pipe (not to scale).

dynamic viscosity of the fluid. For example, when the external flow system of Fig. 4.6 is considered, v is the free stream velocity and D is the distance from the leading edge to the velocity profile; for the internal flow system of Fig. 4.7, the corresponding parameters are mean fluid velocity and hydraulic diameter D_h. *Hydraulic Diameter* is defined as four times the cross-sectional area of the conduit divided by the wetted perimeter of the conduit:

$$D_h \equiv 4\left(\frac{\text{cross-sectional area}}{\text{wetted perimeter}}\right) \tag{4.4}$$

Note that D_h for a completely filled round tube is the tube diameter.

The numerator terms of Re are related to inertial forces while the denominator term is related to the viscous forces. Consequently, Re can be thought of as the ratio of inertial to viscous forces. When Re is small, the viscous forces are large compared with the inertial forces, and so the flow is laminar; a converse relationship results in turbulent flow.

In practice, the transition process from laminar to turbulent flow covers a wide range of Reynolds numbers. It is customary, however, to assume that laminar flow ceases when the following conditions exist:

$$\text{External flow} \quad Re > 5 \times 10^5 \tag{4.5}$$

$$\text{Internal flow} \quad Re > 2300 \tag{4.6}$$

Experiments have shown that the hydrodynamic entry length (L) of Fig. 4.7 is a function of the Reynolds number and pipe diameter (D):

$$\frac{L}{D} \simeq 0.06\ Re \tag{4.7}$$

Assuming that the limit of laminar flow is Re = 2300,

$$L \simeq (0.06)(2300)D = 138\ D$$

That is, a length of 138 pipe diameters is required before internal laminar flow is fully developed. Other experiments indicate that for turbulent flow a length of less than 40 diameters is required.

Prandtl Number

The Prandtl number Pr is a nondimensional grouping of fluid transport properties:

$$Pr \equiv \frac{c\mu}{k} \tag{4.8}$$

where c is the specific heat and k is the thermal conductivity. The number can be expressed in another form by multiplying Eq. (4.8) through by density ρ and recalling that kinematic viscosity $\nu = \mu/\rho$. Also, another property called thermal diffusivity α can be introduced:

$$\alpha \equiv \frac{c\rho}{k} \tag{4.9}$$

Diffusivity of any type is the rate at which an effect or a condition permeates a medium. Therefore, ν is the diffusivity for momentum or velocity and α is the diffusivity for heat or temperature. Consequently,

$$Pr = \frac{\nu}{\alpha} = \frac{\text{kinematic viscosity}}{\text{thermal diffusivity}} \tag{4.10}$$

and represents a measure of how the velocity and temperature profiles of Figs. 4.4 and 4.5 develop relative to one another. If Pr = 1.0, the diffusivities are equal and the profiles will develop at the same rate. If Pr > 1.0, the velocity profile will develop faster than the temperature profile, and if Pr < 1.0, the opposite will happen.

Since all the elements that make up the Prandtl number are thermophysical properties, it is important that temperature be carefully specified. Figure 4.8 shows ranges of Pr for several substances.

Hagen-Poiseuille Equation

For fully developed laminar flow through a tube of small diameter, a special relationship exists for predicting the volume discharge rate Q. The relationship is called the Hagen-Poiseuille equation:

$$Q = \frac{\Delta p}{\ell}\frac{\pi r^4}{8\mu} \tag{4.11}$$

In this equation $\Delta p/\ell$ represents the pressure gradient over a length ℓ and r is the inside radius of the tube as shown in Fig. 4.9. The velocity gradient for fully developed flow in a tube is parabolic. Equation (4.11) agrees very well with measurements made for flow in a capillary, but the error increases rapidly as the tube is made larger.

Stokes' Law

When an ideal fluid flows past a sphere, the streamline pattern around the sphere is perfectly symmetrical as shown in Fig. 4.10a. This idealistic condition results because viscosity is neglected. In such a model the upstream and downstream pressure gradients are also symmetrical, and the resultant force on the sphere is zero. Consequently, there is no resistance to motion between the sphere and an inviscid fluid; this result is known as d'Alembert's paradox.

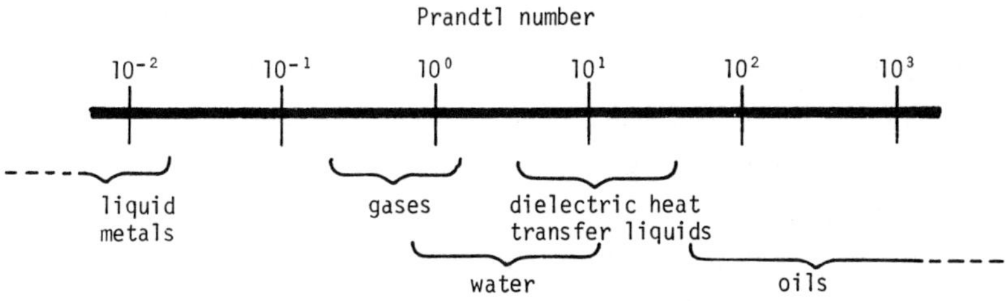

FIGURE 4.8 Spectrum of Prandtl numbers.

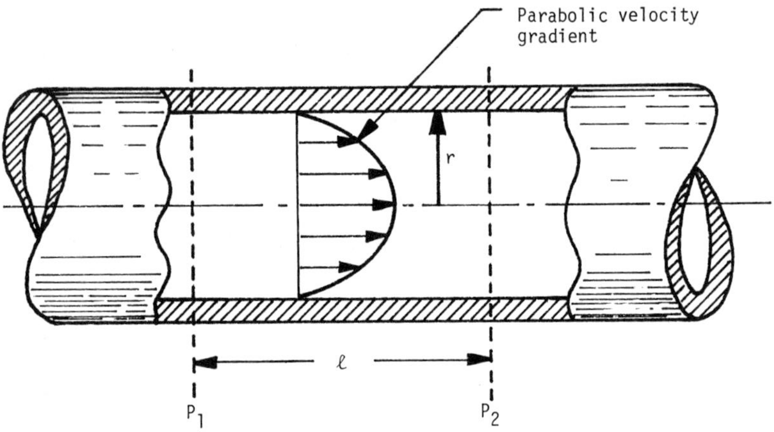

FIGURE 4.9 Model used to develop the Hagen-Poiseuille equation [Eq. (4.11)] for fully developed laminar flow through tubes of small diameter.

If a real, viscous fluid flows past a sphere, a boundary layer will form which will be laminar or turbulent depending on whether the Reynolds number is low or high. In either case the fluid will separate from the sphere at some point along the surface of the sphere. Past the separation point a system of eddies, called the wake, is formed. Pressure within the wake is considerably lower than the static pressure of the undisturbed stream, and it is this decreased pressure field that causes the drag or resistance on the sphere.

The nature of a laminar boundary layer results in less resistance to flow than a turbulent layer, but the laminar layer is much less stable. The laminar layer will always separate before a turbulent layer. For this reason designers often resort to artificial means to cause an early transition from laminar flow to turbulent flow. For example, Fig. 4.11 illustrates schematically the result of placing a thin wire ring over the leading edge of a sphere. Compare this figure with Fig. 4.10 and note the diminished wake. A practical application of this technique is the dimple pattern on a golfball. The dimples tend to change the boundary layer from laminar to turbulent, thereby reducing the drag and allowing the ball to travel a longer distance.

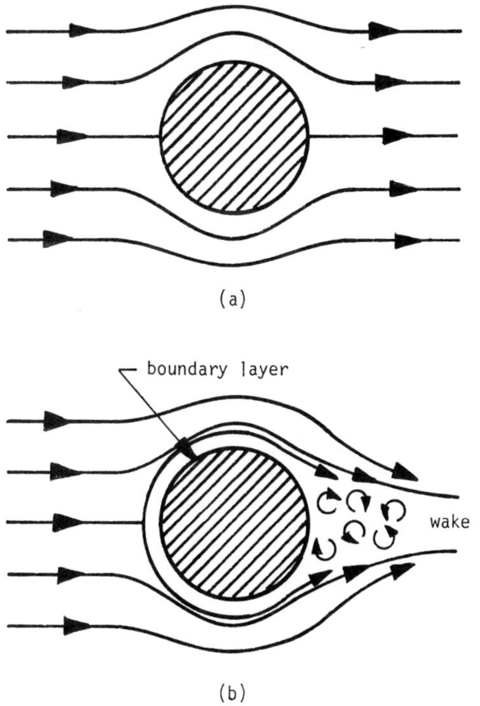

FIGURE 4.10 Fluid flow past a cylinder: (a) inviscid flow and (b) viscid flow.

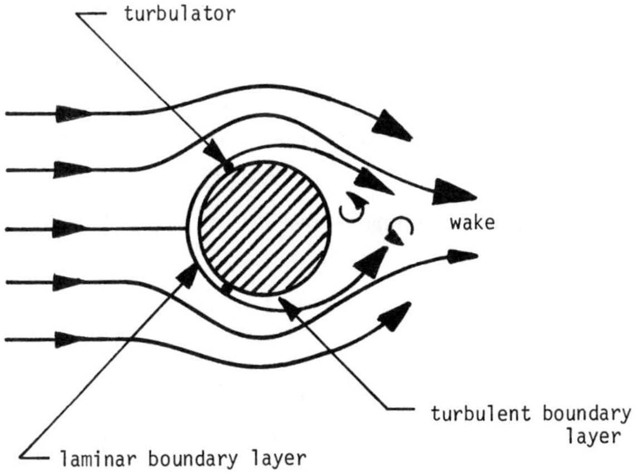

FIGURE 4.11 Transition from laminar to turbulent flow past a cylinder produced by a turbulator.

For low values of Reynolds number ($Re < 1.0$) the drag force F on a sphere moving through a viscous fluid can be predicted by use of Stokes' law:

$$F = 3\pi\mu Dv \tag{4.12}$$

where μ is the dynamic viscosity, D is the sphere diameter, and v is the velocity.

The practical consequence of Stokes' law is that a sphere falling through a viscous liquid will attain a terminal velocity v_t. This velocity can be found by noting that the buoyant force plus the drag force must just equal the weight of the sphere:

$$\frac{4}{3}\pi\frac{D^3}{8}w_f + 3\pi\mu Dv = \frac{4}{3}\pi\frac{D^3}{8}w_s$$

In this relation the subscripts f and s refer to the specific weight of the fluid and the sphere, respectively. Solving for the terminal velocity:

$$v_t = \frac{D^2}{18\mu}(w_s - w_f) \tag{4.13}$$

Bodies of all shapes experience the viscous drag force and all have a terminal velocity, but these values are readily calculable for only a sphere.

4.8 CONCLUSION

In this chapter the very important property of viscosity is covered. The qualitative concept of viscosity is described and methods used to predict quantitative effects are presented. Calculations in the three most important systems of units are shown. Several practical systems used in specifying the viscosity effects of petroleum are outlined. Three types of nonnewtonian fluids are classified: plastic, pseudoplastic, and dilatant.

The most significant engineering consequences of viscosity are its effects on fluid flow, and several of these effects are reviewed, including boundary layers, Reynold and Prandtl numbers, the Hagen-Poiseuille equations, and Stokes' law.

SOLVED PROBLEMS

PROBLEM 1: At 68°F (20°C) the experimental value of absolute viscosity for water is 1 cp. What is the equivalent value in (a) SI units, (b) British Engineering units, and (c) reyns?

Solution: Use conversion units from Table 4.1.

(a) $$1.0 \text{ cp} = 1.0 \times 10^{-2} \text{ poise}$$

$$\frac{1.0 \times 10^{-2}}{10} = 1.0 \times 10^{-3} \text{ Pa s}$$

(b) $$\frac{1.0 \times 10^{-2} \text{ poise}}{4.788 \times 10^{2}} = 2.089 \times 10^{-3} \text{ lbf s/ft}^2$$

$$(2.089 \times 10^{-3} \text{ lbf s/ft}^2)(32.2 \text{ lbm ft/lbf s}^2) = 6.73 \times 10^{-2} \text{ lbm/ft s}$$

(c) $$(2.089 \times 10^{-3} \text{ lbf s/ft}^2)(\text{ft}^2/144 \text{ in}^2) = 1.45 \times 10^{-5} \text{ reyn}$$

PROBLEM 2: Determine the kinematic viscosity of water at 68°F in (a) SI units and (b) British units.

Solution: Use conversion units from Table 4.1.
At 68°F

$$\rho_{H_2O} = 1.0 \text{ gm/cm}^3$$

$$\mu_{H_2O} = 1.0 \times 10^{-2} \text{ gm/cm s}$$

$$\nu \equiv \frac{\mu}{\rho} = \frac{1.0 \times 10^{-2} \text{ gm/cm s}}{1.0 \text{ gm/cm}^3} = 10^{-2} \text{ cm}^2/\text{s}$$

(a) $$(10^{-2} \text{ cm}^2/\text{s})(\text{m}^2/10^4 \text{ cm}^2) = 10^{-6} \text{ m}^2/\text{s}$$

(b) $$(10^{-6} \text{ m}^2/\text{s})(10.76) = 1.076 \times 10^{-5} \text{ ft/s}$$

PROBLEM 3: Determine the kinematic viscosity of a 40 SUS oil.

Solution: Use conversion units from Table 4.2:

$$\nu = (0.0022)(40) - \frac{1.80}{40} = 0.043 \text{ stoke}$$

PROBLEM 4: Air at 70°F flows through a duct at 50 fps. Calculate the Reynolds number if the duct is (a) 10 in in diameter (b) 10 in square.

Solution:

$$\mathrm{Re} \equiv \frac{\rho v D}{\mu} = \frac{vD}{\nu}$$

where

v = 50 fps

$\nu = 1.6 \times 10^{-4}$ ft^2/s (Ref. 2)

D = hydraulic diameter

$$D \equiv 4\left(\frac{\text{cross-sectional area}}{\text{wetted perimeter}}\right)$$

(a) cross-sectional area = $\pi(10)^2/4$ in^2, wetted perimeter = 10π in:

$$\frac{\pi(10)^2/4}{10\,\pi} = \frac{10}{4} \quad \text{and} \quad D = 4\left(\frac{10}{4}\right) = 10 \text{ in} = \frac{5}{6} \text{ ft}$$

$$\therefore \quad \mathrm{Re} = \frac{(50)(5/6)}{(1.6 \times 10^{-4})} = 260\ 000$$

(b) cross-sectional area = 10^2 in^2, wetted perimeter = (4)(10) in:

$$\frac{10^2}{(4)(10)} = \frac{10}{4} \quad \text{and} \quad D = \frac{5}{6} \text{ ft}$$

$$\therefore \quad \mathrm{Re} = 260\ 000$$

PROBLEM 5: A fluid with a viscosity of 1.75×10^{-3} Pa s flows through a tube of 0.5 mm diameter. If the pressure drop across a 1 m length of the tube is 1.0 MPa, and the flow is fully developed and laminar, what is the flow rate?

Solution: Since the flow is fully developed and laminar, the Hagen-Poiseuille equation can be applied:

$$Q = \frac{\Delta p\ \pi r^4}{\ell 8 \mu}$$

where

Q = flow rate

Δp = 1.0 MPa = 10^6 N/m^2

r = 0.25 mm = 2.5×10^{-4} m

ℓ = 1.0 m

$\mu = 1.75 \times 10^{-3}$ Pa s = 1.75×10^{-3} Ns/m^2

$$Q = \frac{(10^6)(\pi)(2.5 \times 10^{-4})^4}{(1.0)(8)(1.75 \times 10^{-3}\ \mathrm{Ns/m^2})} = 8.77 \times 10^{-7}\ \mathrm{m^3/s}$$

EXERCISES

1. If the viscosity of a liquid is 3 centipoises, express the viscosity in (a) SI units and (b) British absolute units.
2. If the liquid in problem 1 has a specific gravity of 1.5, determine the kinematic viscosity in (a) SI units and (b) British units.
3. Determine the kinematic viscosity of 40 SFS oil.
4. Use a handbook to determine the dynamic and kinematic viscosities of air and water at 20°C. Comment on the results.
5. What are the effects of pressure changes on (a) absolute viscosity and (b) kinematic viscosity?
6. What is the essential difference between Newtonian and nonnewtonian fluids?
7. Define a thixotropic fluid.
8. Why is the concept of an inviscid fluid useful?
9. Water at 70°F flows through a duct at 50 fps. Calculate the Reynolds number if the duct is (a) 1 ft in diameter and (b) a 6 × 12 in rectangle.
10. A fluid with a viscosity of 1.6 poises flows through a tube 0.6 mm in diameter. If the pressure drop across a 1 m length is 1.0 $\mathrm{N/mm^2}$, and if the flow is fully developed laminar, what is the flow rate?

REFERENCES

1. E. R. G. Eckert and R. M. Drake, *Heat and Mass Transfer*, 2d ed., McGraw-Hill, New York, 1959.
2. F. Kreith, *Principles of Heat Transfer*, 2d ed., appendix III, table A-3, International Textbook Co., Scranton, Pa., 1967.
3. O. W. Eshbach, *Handbook of Engineering Fundamentals*, pp. 1-140 to 1-141 and 6-06 to 6-09, Wiley, New York, 1952.
4. V. L. Streeter and E. B. Wylie, *Fluid Mechanics*, appendix C, McGraw-Hill, New York, 1975.

5. *Handbook of Chemistry and Physics*, pp. F-34 to F-46, Chemical Rubber Co., Cleveland, Ohio, 1975.

BIBLIOGRAPHY

Fox, R. W., and A. T. McDonald, *Introduction to Fluid Mechanics*, pp. 329-403, John Wiley, New York, 1978.

Robinson, J. L., *Basic Fluid Mechanics*, pp. 117-137, McGraw-Hill, New York, 1963.

Schlichting, H., *Boundary Layer Theory*, 6th ed., pp. 1-36, McGraw-Hill, New York, 1968.

Vennard, J. K., *Elementary Fluid Mechanics*, 4th ed., pp. 222-255, John Wiley, New York, 1962.

5

SURFACE TENSION

5.1 INTRODUCTION

A liquid in a container will assume a shape that exhibits the smallest possible free surface area. In this sense it acts like an elastic membrane under tension and the property responsible for the phenomenon is called surface tension σ. In Fig. 5.1 a schematic illustration shows that molecule A in the interior of the liquid is attracted equally in all directions by the surrounding molecules (cohesive forces); surface molecule B is subjected to cohesive forces from below the surface and to adhesive forces in the vapor above the surface. The net force on B is normal to the surface and directed toward the

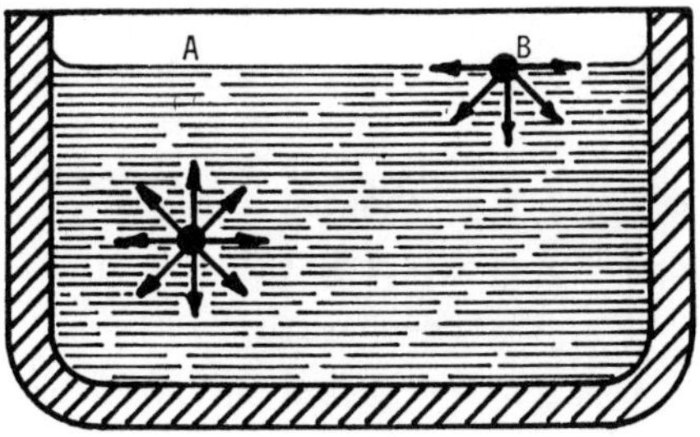

FIGURE 5.1 Cohesive and adhesive forces on molecules A and B in a container of liquid.

interior of the liquid. Consequently, because molecules at or near the surface are attracted toward the interior, there is a tendency for the liquid surface to become as small as possible. It is this characteristic that causes droplets of water falling through the air to form themselves into almost perfect spheres, since a sphere has the smallest surface area for a given volume.

5.2 UNITS AND DIMENSIONS

The dimension of surface tension can be expressed by considering the characteristics of the minimum surface area that forms in accordance with the least energy principle. The only way the surface area can be increased is for some interior molecules to move to the surface, and since these molecules are at a lower energy level than the surface molecules, an expenditure of energy is required. The surface tension can then be considered as the work required to increase the surface area:

$$\text{Surface tension} \equiv \frac{\text{work}}{\text{increase in surface area}}$$

$$[\sigma] = \frac{[F][L]}{[L^2]} = \frac{[F]}{[L]}$$

Consequently, the dimensions for surface tension are force per length. The units in the SI system are N/m, and in the British system they are lbf/ft. Most values of surface tension are considerably less than 1 N/m, so a modified unit, mN/m (= 10^{-3} N/m), is commonly used for work in the metric system. This unit equals 1 dyn/cm, which is the cgs unit.

Values of surface tension are generally reported for a substance in contact with its own vapor or in contact with air. The property is thermophysical and decreases with temperature. Table 5.1 shows typical values of surface tension for several substances.

5.3 CONTACT ANGLE

The net force on molecule B in Fig. 5.1 results from the differences in the cohesive and adhesive forces which act at the boundary between

TABLE 5.1 Typical Values of Surface Tension for Several Substances

Substance	In contact with	Temp. (°C)	mN/m
Water	Air	0	75.6
		20	72.8
		40	69.6
		60	66.2
		80	62.6
		100	58.9
Carbon Tetrachloride (CCl_4)	Vapor	20	27.0
		100	17.3
		200	6.5
Ethyl Alcohol (C_2H_6O)	Vapor	10	23.6
		20	22.8
		30	21.9
Mercury	Air	20	514
	Vacuum	20	484
Fluorinated Hydrocarbons	Air	25	11-20

[mN/m] x 10^{-3} = [dyn/cm] x 10^{-3} = [N/m]

[N/m] x 0.069 = [lbf/ft]

the vapor and the liquid. It is evident that, in the general case, three different interfaces will be formed as shown in Fig. 5.2, and three different values of surface tension will exist:

σ_{SL} = surface tension of the solid-liquid film
σ_{SV} = surface tension of the solid-vapor film
σ_{LV} = surface tension of the liquid-vapor film

The diagrams in Fig. 5.3 show that the characteristic values of σ for different combinations of liquids and solids result in characteristic shapes of the free surface. The three films meet at the wall and this junction is shown in each of the diagrams. The junction is assumed to be of unit thickness in the plane of the paper and is acted on by four forces; three of these forces are due to surface tension and the fourth is the adhesive force F_a between the junction and the solid wall. If conditions of equilibrium are applied to the junction:

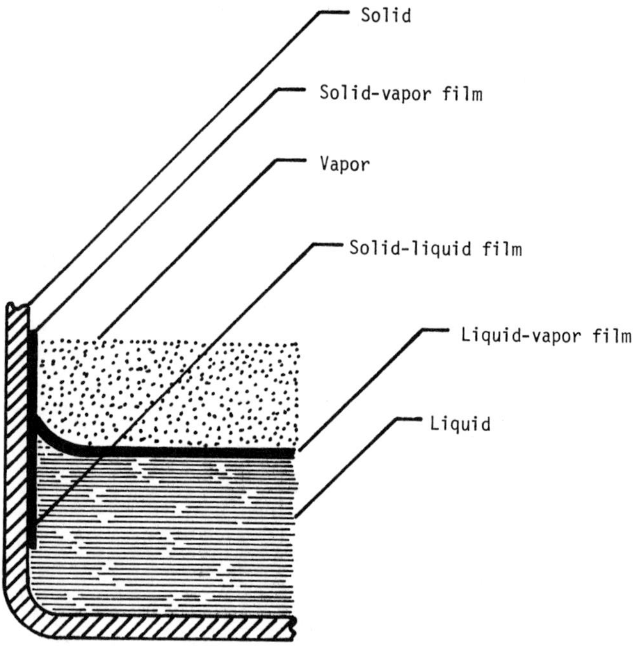

FIGURE 5.2 Three types of surface films.

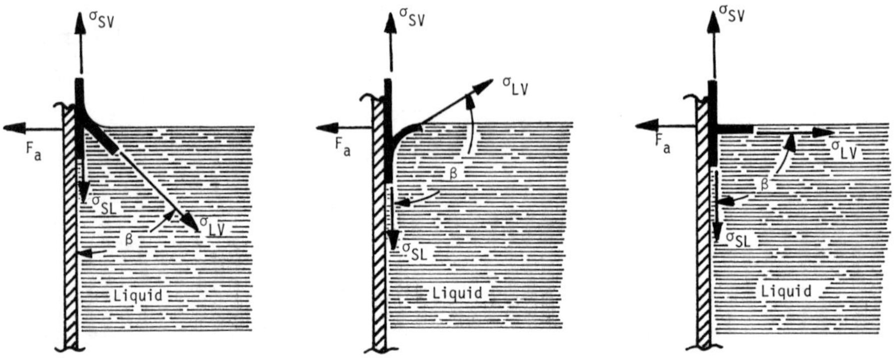

FIGURE 5.3 Characteristic shapes of free liquid surfaces.

$$\Sigma F_x = \sigma_{LV} \sin \beta - F_a = 0$$

$$\Sigma F_y = \sigma_{SV} - \sigma_{SL} - \sigma_{LV} \cos \beta = 0$$

from which

$$F_a = \sigma_{LV} \sin \beta \tag{5.1}$$

$$\sigma_{SV} - \sigma_{SL} = \sigma_{LV} \cos \beta \tag{5.2}$$

The angle β is called the *contact angle* and is customarily measured from the solid surface, through the liquid, to a tangent to the liquid free surface.

Equation (5.1) enables one to calculate the adhesive force if σ_{LV} and β are shown. Equation (5.2) indicates that the contact angle β depends on the difference between σ_{SV} and σ_{SL}.

The following relationships can be deduced from Eq. (5.2):

If $\sigma_{SV} > \sigma_{SL}$, then $\cos \beta > 0$, and $0 < \beta < 90°$

If $\sigma_{SV} < \sigma_{SL}$, then $\cos \beta < 0$, and $90° < \beta < 180°$

If $\sigma_{SV} = \sigma_{SL}$, then $\cos \beta = 0$, and $\beta = 90°$

In the first case the liquid surface curves upward on the wall and the liquid is said to wet the wall; in the second case the surface is depressed and the liquid does not wet the wall; if the last condition is obtained, the liquid surface will be normal to the wall.

5.4 PRESSURE DIFFERENCE ACROSS A FILM SURFACE

Figure 5.4 shows a small element of the surface of a film formed by surface tension. In general such a film will have a double curvature of radii R_1 and R_2. The only forces acting on the element are a result of the difference between internal and external pressures p_i and p_o, respectively, and the surface tension σ. A relation between the forces can be shown by assuming static equilibrium and letting the summation of forces normal to the element be zero:

$$(p_i - p_o)dx\, dy = 2\sigma\, dy \sin \theta + 2\sigma\, dx \sin \phi$$

but

$$\sin \theta = \frac{dx}{2R_1} \quad \text{and} \quad \sin \phi = \frac{dy}{2R_2}$$

$$p_i - p_o = \sigma \left(\frac{1}{R_1} + \frac{1}{R_2}\right) \tag{5.3}$$

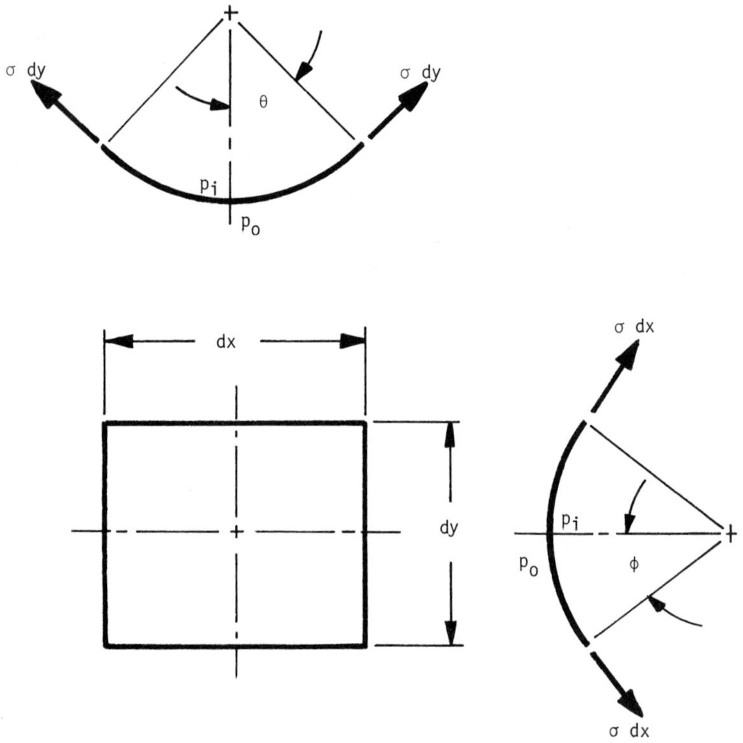

FIGURE 5.4 Liquid element of double curvature under surface tension forces.

Equation (5.3) states that the pressure differential across a film is a function of surface tension and the curvature of the film.

Any surface under tension will contract until it occupies the minimum area consistent with the boundaries of the surface and the pressure differential. As a consequence, freely falling water drops, soap bubbles, and vapor bubbles formed in a nucleate boiling process all tend to assume spherical shapes under the action of surface tension. The relation between pressure forces and surface tension forces for such a condition can be analyzed by considering a hemisphere of radius R as shown in Fig. 5.5. The surface tension acts along the perimeter of the hemisphere and the components of the pressure forces

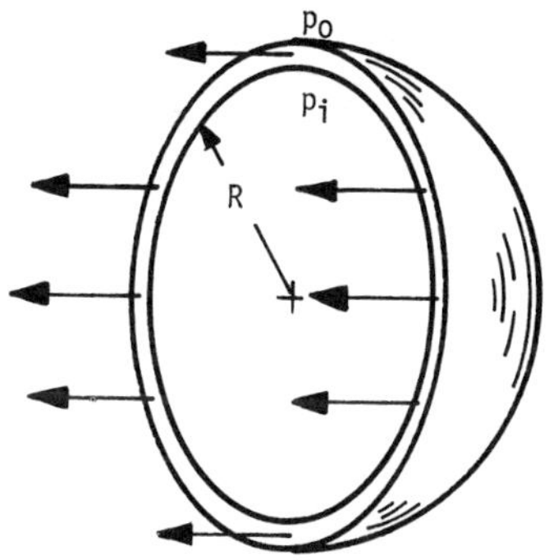

FIGURE 5.5 Liquid hemisphere.

can be considered to act on the projected areas. The vertical forces can be identified as

Surface tension force = $\sigma 2\pi R$

Internal pressure force = $p_i \pi R^2$

External pressure force = $p_o \pi R^2$

If static equilibrium is assumed, the sum of the vertical forces is zero:

$$p_i \pi R^2 = p_o \pi R^2 + \sigma 2\pi R$$

from which

$$p_i - p_o = \frac{2\sigma}{R} \qquad (5.4)$$

Equation (5.4) states that the internal pressure of a spherical bubble exceeds the external pressure by an amount $2\sigma/R$. Actually, it is a special case of Eq. (5.3) when $R_1 = R_2$.

According to Eq. (5.4) the internal pressure p_i will decrease as the diameter of the bubble is increased. Experimental verification of this conclusion can be obtained by means of the apparatus shown in Fig. 5.6. Two bubbles A and B of different size are formed and isolated from one another by closing the appropriate valves. When the bubbles are connected by opening these valves, the larger bubble

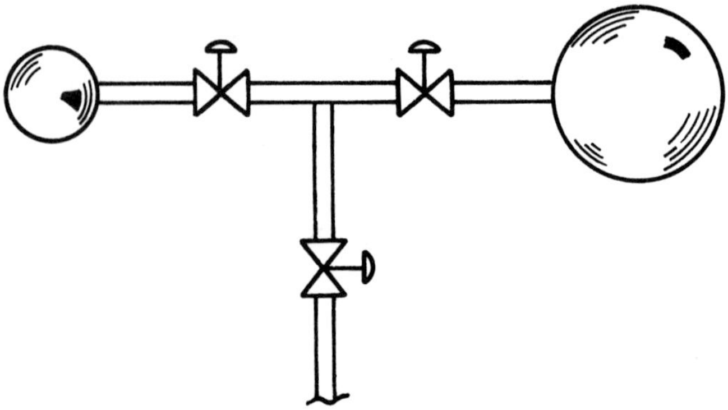

FIGURE 5.6 Bubble pressure test apparatus.

will be inflated by the deflating, smaller bubble. This is conclusive evidence that the pressure in the smaller bubble is greater than that in the larger one.

5.5 CAPILLARITY

The most familiar effect of surface tension is the tendency for liquids to move through passageways or ducts of small cross section. The name applied to the phenomenon is capillarity which is derived from the Latin word for hair or hairlike.

The effects of surface tension are evidenced in many common materials. Consider, for example, trying to write with ink on a paper towel. The ink will diffuse through the pores in the paper and the character will be disfigured. On the other hand, the paper towel can be used to absorb water, whereas a bond paper whose pores have been closed will not readily absorb moisture. A dry sponge will not absorb water because the pores are generally closed, but when the sponge is wet and squeezed out, the pores are open and large quantities of water can be absorbed.

Capillarity effects in the soil are used by dirt farmers. After a heavy rain the soil is saturated to a considerable depth. When the sun shines again, the water close to the surface evaporates and the capillarity in the pores of the soil brings more water to the surface, where it in turn evaporates. If, however, the surface soil is cultivated so that it is loosened or pulverized, the pores become too large for an effective capillary action and the evaporation is reduced. Meanwhile, water is still supplied to the roots of the plant by capillarity where it can be used to help plant growth.

The heat pipe also makes use of the principle of capillarity. Figure 5.7 is a cross section of a typical heat pipe. The working liquid is evaporated in the boiler section by radiated, convected, or conducted heat. The resulting vapor moves through the center of the pipe, through the action of a pressure differential, to the condenser section where heat is removed by radiation, convection, or conduction. The wick then provides a path for the condensed liquid to return to the boiler section under the action of capillary forces.

5.6 MENISCI

For any system in which a measurement is made by observing the liquid level in a small tube, the effects of surface tension are particularly important. The liquid will either rise or fall in the tube, and a curved liquid surface will be formed which is called a *meniscus*. It

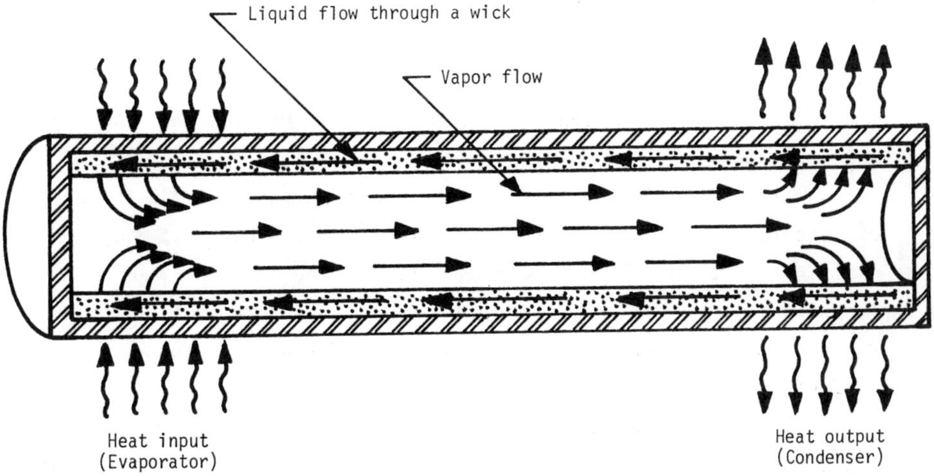

FIGURE 5.7 Heat pipe.

was shown previously that a liquid which wets the tube will rise, and the contact angle β will be less than 90°; if the liquid is nonwetting, the meniscus will be depressed and β will exceed 90°.

Figure 5.8 shows a wetting and a nonwetting liquid in a tube of diameter D. A cylinder of height h and diameter D can be isolated as a free body, and the conditions of equilibrium can be applied to it. Surface tension forces will act around the periphery of the meniscus and the total upward force can be expressed as

$$\text{Total upward force} = \pi D\sigma \cos \beta$$

The total downward force is the weight of the liquid cylinder, which can be expressed as a function of the mass density ρ and the acceleration of gravity:

$$\text{Total downward force} = \rho g h \frac{\pi D^2}{4}$$

and, since the forces are in equilibrium,

$$\pi D\sigma \cos \beta = \rho g h \frac{\pi D^2}{4}$$

$$h = \frac{4\sigma \cos \beta}{\rho g D} \tag{5.5}$$

Equation (5.5) expresses the height that a liquid will rise in a capillary tube. As an exercise, show that the same equation can be used to determine how far a liquid will be depressed. Keep in mind that the liquid will rise when $0 \le \beta \le 90°$ and fall when $90° \le \beta \le 180°$. The surface tension σ is assumed to be that between the liquid and its vapor.

Equation (5.5) can also be derived by considering the pressure differential across a film surface. For example, Fig. 5.9 shows the meniscus of radius R for a wetting liquid. The pressure above the meniscus is atmospheric pressure p_a and the pressure just below the surface is p. Recall that the pressure differential across a film surface was given by Eq. (5.4).

When this equation is applied to Fig. 5.9, $p_i = p_a$ and $p_o = p$; also, the radius R can be expressed in terms of the tube diameter D:

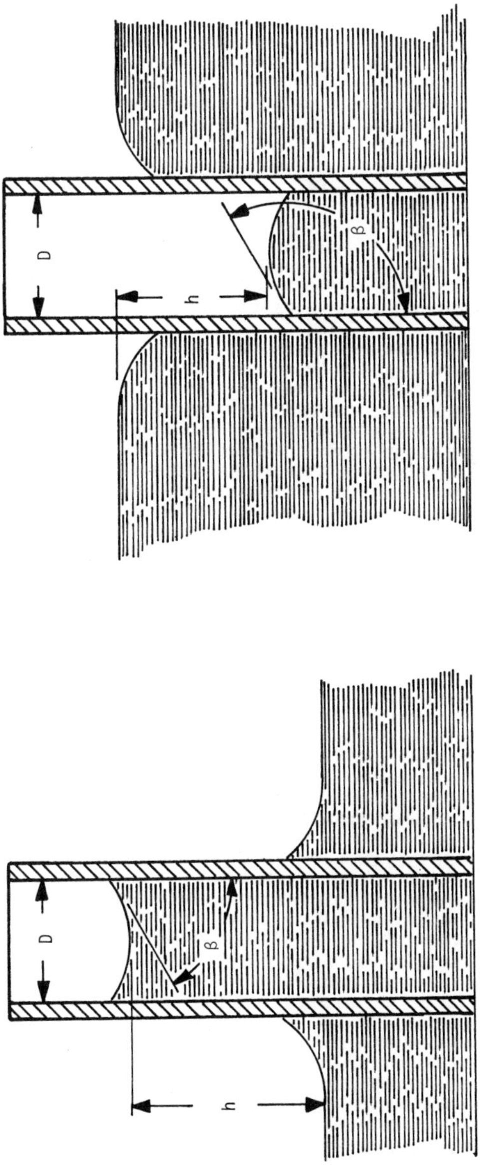

FIGURE 5.8 Menisci formations through the action of surface tension: (a) wetting liquid and (b) nonwetting liquid.

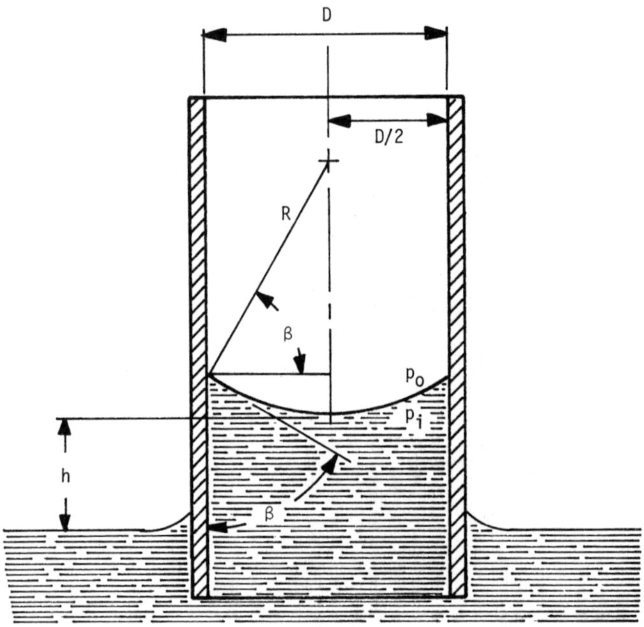

FIGURE 5.9 Pressure differential across a meniscus.

$$R = \frac{D}{2 \cos \beta}$$

and applying Eq. (5.4)

$$p_a - p = \frac{4\sigma \cos \beta}{D}$$

The pressure in the column at a point whose elevation is the same as the bulk liquid surface is also atmospheric pressure p_a. Therefore, the pressure p at a height h can be expressed as

$$p = p_a - \rho g h$$

Pressure p in the latter two relationships can be equated so that the following relationship is obtained:

$$\rho g h = \frac{4\sigma \cos \beta}{D}$$

which gives the same result as previously obtained:

$$h = \frac{4\sigma \cos \beta}{\rho g D} \tag{5.5}$$

Equation (5.5) applies to pure liquids and perfectly clean solid surfaces, conditions which are not normally found in practice. As a consequence, experimental values of h will be somewhat less than indicated by the equation; use of the equation will, however, result in conservative estimates of the meniscus error. Since h is inversely proportional to the tube diameter D, relatively large diameters should be used to avoid meniscus errors. If $D > 3/8$ in, the rise or fall of the liquid can generally be neglected.

5.7 CONCLUSION

In this chapter the thermophysical property of surface tension is introduced. This is a phenomenon that acts at the interface between a liquid and a gas, or between two immiscible liquids. Surface tension is also responsible for capillary action, by which liquids tend to rise or fall within tubes of small diameter, depending upon whether the liquid wets or does not wet the sides of the tube.

Qualitative and quantitative effects of surface tension are outlined, and the most commonly used units are given. Several important consequences of surface tension are discussed, including pressure differences across films, capillarity, and menisci.

SOLVED PROBLEMS

PROBLEM 1: If the surface tension of a soap bubble is 0.04 N/m, what work is required to increase the diameter from 5 to 15 cm?

Solution:

Work = (surface tension)(increase in area)

The surface area of a sphere is $4\pi r^2$, but there is an inside surface and an outside surface to be increased.

$$\text{Work} = (40 \text{ dyn/cm})(2)(4\pi)[(7.5 \text{ cm})^2 - (2.5 \text{ cm})^2]$$

$$= 5.03 \times 10^4 \text{ dyn cm or erg}$$

PROBLEM 2: By how much does the pressure inside a 0.017 in droplet of water at 20°C exceed the outside pressure:

Solution:

$$p_i - p_o = \frac{2\sigma}{R}$$

where

$$p_i - p_o = \Delta p \text{ (the pressure excess)}$$

$$\sigma = 72.8 \text{ m N/m at } 20°\text{C (from Table 5.1)} = 0.005 \text{ lbf/ft}$$

$$R = 0.017 \text{ in} = 7.08 \times 10^{-4} \text{ ft.}$$

$$\Delta p = \frac{(2)(0.005)}{7.08 \times 10^{-4}} = 14.1 \text{ lbf/ft}^2$$

PROBLEM 3: A 0.4-mm diameter glass tube stands vertically in a dish of mercury at 20°C. Determine the difference between the level of mercury in the dish and in the tube. The specific gravity for mercury is 13.6 and the contact angle with glass is 130°.

Solution: Since Hg does not wet glass, the level in the tube will be depressed a distance h (see Fig. 5.8b).

$$h = \frac{4\sigma \cos \beta}{\rho g D}$$

where

$$\sigma = 514 \text{ dyn/cm}$$

$$\rho = 13.6 \text{ gm/cm}^3$$

$$g = 980 \text{ cm/s}^2$$

$$D = 0.04 \text{ m}$$

$$h = \frac{(4)(514)(\cos 130°)}{(13.6)(980)(0.04)} = -2.48 \text{ cm}$$

The negative sign denotes a depression.

EXERCISES

1. Using the data in Table 5.1, express the surface tension of 60°F water in contact with air in (a) lbf/ft and (b) dyn/cm.

2. If the surface tension of a soap bubble is 0.0025 lbf/ft, what is the work required to increase the diameter from 3 to 5 in?
3. What is a contact angle?
4. If the contact angle is 45°, does a liquid wet a surface?
5. Give an example of a liquid that does not wet a solid surface.
6. Why does a safety razor blade float if it is carefully placed in water? Compare the phenomenon with that which causes a steel ship to float.
7. What is the diameter of a soap bubble if the difference in pressure between the inside and the outside is 10^3 Pa? (Let $\sigma = 0.04$ N/m.)
8. Why does blotting paper dry ink without smearing?
9. An open glass tube with an inside diameter of 0.05 in is placed vertically in a dish of mercury. If the mercury is depressed 0.3 in, determine the surface tension of the mercury. The specific gravity for mercury is 13.6 and the contact angle is 130°.
10. A liquid whose specific gravity is 0.8 rises 50 cm in a capillary tube of 0.04-mm bore and the contact angle is observed to be 20°. What is the surface tension of the liquid?

BIBLIOGRAPHY

Handbook of Chemistry and Physics, 49th ed., pp. F-19 to F-33, Chemical Rubber co., Cleveland, Ohio, 1969.

Robinson, J. L., *Basic Fluid Mechanics*, pp. 7-10, McGraw-Hill, New York, 1963.

Sears, F. W. and M. W. Zemansky, *University Physics*, 3rd ed., pp. 296-308, Addison-Wesley, Reading, Mass., 1964.

Streeter, V. L. and E. B. Wylie, *Fluid Mechanics*, 6th ed., pp. 18-20, McGraw-Hill, New York, 1979.

Vennard, J. K., *Elementary Fluid Mechanics*, 4th ed., pp. 17-20, Wiley, New York, 1962.

6

WORK AND HEAT

6.1 INTRODUCTION

Work and heat are two of the more important concepts used in thermal technology. They are an essential part of the first law of thermodynamics, which is the bedrock foundation for all thermal analysis. Work and heat are different manifestations of the same phenomenon, namely, a transfer of energy. Consequently, a complete understanding of the concepts must be accompanied by an understanding of energy itself.

6.2 CONCEPT OF ENERGY

Energy is a primitive concept and no attempt is made to define it. However, the effects of changes in energy can be observed, measured, and harnessed to perform useful tasks; it is these changes that are identified and quantified.

For the purpose of quantification, energy can be considered as a mathematical abstraction that has no existence apart from a functional relationship to certain variables. For example, kinetic energy (KE) of a given mass (m) is a function of the velocity (v):

$$KE \equiv \frac{mv^2}{2g_c} \tag{6.1}$$

Also, potential energy (PE) is a function of the elevation of a mass above a datum Z and the acceleration of gravity g:

$$PE \equiv \frac{mgZ}{g_c} \tag{6.2}$$

Therefore, kinetic and potential energies may be considered to have no reality apart from the equations shown above. In addition to these two forms of energy there are many other forms, a few of them being electrical, chemical, magnetic, acoustical, luminous, and nuclear.

Although energy cannot be defined unequivocally, three straight-forward statements can be made concerning energy:

All substances contain energy.
Total energy is the sum of all the individual energies.
Energy is conserved.

These statements, in total, make up the first law of thermodynamics. A qualitative statement of the law is the axiom: energy cannot be created or destroyed but only transferred from one form to another. Quantitative statements of the law are embodied in a variety of equations which are presented in Chap. 7.

Internal Energy

In the study of thermal technology it is customary to consider kinetic and potential energy separately and to lump all the other energies into a property called *internal energy*, to which the symbol U is assigned. Internal energy is an extensive property, but it is convenient to work with the intensive version--specific internal energy u--which is internal energy per unit mass. Tables of u for various substances are available in the literature [1-3]. Such listings are possible because internal energy can be considered as one of the independent properties of a pure substance.

Equations for the specific internal energy of a substance between the saturated liquid and saturated vapor states can be written and are analogous to Eqs. (1.3) to (1.5):

$$\begin{aligned} u &= u_f + xu_{fg} \\ &= xu_g + (1 - x)u_f \\ &= u_g - (1 - x)u_{fg} \end{aligned} \tag{6.3}$$

In Eq. (6.3) x is the quality, f refers to liquid, g to vapor, and fg to a state between saturated liquid and saturated vapor.

Flow Energy

Many thermal systems depend upon the action of a fluid being pumped through the system. Whenever the fluid crosses a system boundary there is a certain amount of energy associated with the pumping action required to effect the boundary crossing. This energy is called *flow energy* [4] and it is not given a special symbol but is generally referred to as the product of the pressure and volume of the fluid at the boundary:

$$\text{Flow energy} \equiv PV \tag{6.4}$$

Flow energy is an extensive, thermodynamic property. For convenience, a corresponding intensive property is also defined in terms of specific volume (v):

$$\text{Specific flow energy} \equiv Pv \tag{6.5}$$

Enthalpy

Flow energy is not generally used in thermal calculations; instead, an extensive thermodynamic property called *enthalpy* H is defined as the sum of the internal energy and the flow energy associated with the fluid as it crosses the boundary [4]:

$$H \equiv U + PV \tag{6.6}$$

In addition, an intensive property (specific enthalpy h) is defined as

$$h \equiv u + Pv \tag{6.7}$$

The latter property is the one that is tabulated in the literature [1-3]. In keeping with the procedure used for other thermodynamic properties, specific enthalpy can be expressed in terms of the quality and the properties of saturated liquid and vapor:

$$\begin{aligned} h &= h_f + xh_{fg} \\ &= xh_g + (1 - x)h_f \\ &= h_g - (1 - x)h_{fg} \end{aligned} \tag{6.8}$$

Since enthalpy depends on flow energy, the concept is valid only when a fluid crosses a system boundary; that is, the property of enthalpy is associated only with an open system analysis. For example, unless air is being pumped into or out of a room, the air does not have a property of enthalpy since the room is a closed system.

The terms "total heat" or "heat content" are sometimes used as synonyms for enthalpy. Such usage tends to distort the facts since it is not precise to say that a substance contains "heat." The terms are a throwback to the days when a caloric theory was in vogue.

Prior to the nineteenth century, the caloric theory was widely accepted by the scientific community. According to this theory, there was an undefined substance called *caloric* that could be transferred from one body at a particular temperature to another body at a lower temperature. The theory was consistent with the observed effect that two bodies which were originally at different temperatures would reach the same temperature when placed together for some period of time.

The first conclusive evidence that the theory was invalid was reported by Benjamin Thompson (Count Rumford), an American born physicist, who achieved fame in England. Thompson was engaged in boring cannon for the Bavarian government when he made the observation. It was customary to cool the cannon during the boring operation by flooding it with water. The water would tend to boil and, thus, had to be constantly replenished. According to the caloric theory, the boiling was a consequence of caloric being supplied to the water from the metal shavings. It was believed that a material could release caloric at a faster rate when the material was finely divided, as it would be in the form of shavings.

Thompson observed, however, that the boiling occurred even when the tool was so dull that it did little cutting. As a matter of fact, the boiling became more vigorous as the tool became duller. The effect was contrary to the caloric theory and was reported by Thompson in a paper.

The evidence was subsequently used by several researchers to develop the theory that heat is actually a form of energy and not a substance. The idea was expanded and eventually appeared as the first law of thermodynamics. Joule made careful measurements with falling weights that turned a paddle, which then stirred water in a container. This action raised the temperature of the water, indicating that heat had been transferred to the water from the mechanical work done by the paddle. Joule's measurements have been refined by the use of more accurate equipment and are now accepted as the *mechanical equivalent of heat* J (see the following section).

Units for Energy

When heat is applied to a substance, the temperature of the substance rises. It is not surprising that this temperature-raising effect was the earliest method used to establish a unit for heat measurement. The *calorie* (cal) was originally defined as the quantity of heat required to raise the temperature of water through one celsius degree, and the *British Thermal Unit* (Btu) was defined as the quantity of heat required to raise the temperature of water through one fahrenheit degree. These units and definitions are remnants of the old caloric theory, but the units have been retained and still find wide usage. By international agreement, the cal is now defined as a multiple of the joule; the *joule*, in turn, is defined as the energy transferred when one newton of force causes a displacement of one meter in the direction of the force:

$$1.0 \text{ cal} \equiv 4.186 \text{ joule}$$

The Btu is defined in terms of the calorie through the relation

$$1.0 \text{ cal/gm } ^\circ\text{C} = 1.0 \text{ Btu/lbm } ^\circ\text{F}$$

so that 1.0 Btu = 251.996 joule = 778.26 ft lbf.

The mechanical equivalent of heat J can be expressed in terms of the units defined above:

$$J = 4.186 \text{ joule/cal} = 778.26 \text{ ft lbf/Btu} \tag{6.9}$$

Care must be exercised when using Eq. (6.9); the mechanical equivalent of heat is designated J, which is also the designated symbol for a joule.

Another common energy unit is the large calorie or kilocalorie (kcal), which is equal to 1000 ordinary calories. The kcal is generally used to designate the energy content in various types of food. In the cgs system the unit of energy is the *erg*, defined as the energy transferred when one dyne of force is displaced one centimeter in the direction of the force:

$$1 \text{ erg} = 10^{-7} \text{ joule}$$

For workers in the electrical field it is sometimes convenient to express energy in terms of *electron volt* (eV), defined as the energy required to move one electron through a potential difference of one volt:

$$1 \text{ eV} = 1.602 \times 10^{-19} \text{ joule}$$

The electron volt is a very small amount of energy. When high energy electric fields are being analyzed, the kilo-, mega-, giga-, and tera-electron volts are often used; these units represent 10^3, 10^6, 10^9, and 10^{12} eV, respectively.

For many engineering calculations it is more appropriate to consider the time rate of energy transfer rather than energy transfer. Units for energy rate can generally be found by dividing energy units by time; however, the concept is so widely used that special units such as the watt (W) and the horsepower (hp) have been developed:

$$1.0 \text{ W} \equiv 1.0 \text{ J/s}$$
$$1.0 \text{ hp} \equiv 550 \text{ ft lbf/s}$$
$$1.0 \text{ hp} = 746 \text{ W}$$

Anyone engaged in thermal calculations--or any engineering calculation for that matter--should have ready access to unit conversion tables. In most instances an analyst will find it necessary to consult a number of different sources to obtain all the properties needed in an analysis. It is highly unlikely that all the sources will use the same kinds of units, and so a certain amount of conversion will be necessary. References 5-7 are three sets of tables that are reasonably complete and still retain a certain amount of portability.

6.3 CONCEPT OF WORK

Work (W) is one form of energy transfer. Classically, the work done between states 1 and 2 is defined in terms of a force (F) acting through a displacement (dx):

$$_1W_2 \equiv \int_1^2 F \cdot dx \tag{6.10}$$

In this relationship F is the component of force acting in the same direction as the displacement dx. Work is a dot product of two vectors, thus it is a scalar quantity. A commonly accepted sign convention designates work as positive when it is transferred from a system and negative when it is transferred to a system.

As used in thermodynamics, the word "work" has some unique characteristics not normally associated with the everyday meaning of the word. The exertion of effort is a necessary, but not sufficient, criterion for thermodynamic work to be done. For example, if one pushes against a large stone, a significant amount of effort can be exerted but, as long as the stone does not move, no work is done. Also, if one applies a force to the stone at 45° to the ground, and if the stone moves horizontally, only the horizontal component of the force is said to do work. Although, certainly, the vertical component also requires an expenditure of energy.

To solve Eq. (6.10) it is necessary that the functional relationship between force and displacement be known for each point of the process from state 1 to state 2. This requires the assumption of a quasistatic process so that there will be an infinitesimal deviation from equilibrium at each point (see Chap. 1).

Work can be done in a variety of ways, including rotating shafts, moving pistons, expanding membranes, stretching wires, electrical flow, magnetization, and chemical action. In all cases it is possible to show that the external effect on the system is equivalent to the raising or lowering of a weight. This is consistent with Eq. (6.10) as the fundamental definition of work. For example, consider the simple case of a battery-driven motor shown in Fig. 6.1. Initially, chemical action between the anode and cathode in the battery is converted into a flow of electricity which drives the motor. The motor, in turn, causes a shaft to rotate. By selecting the battery and the motor as a closed system, it is obvious that the final effect could be the raising or lowering of a weight.

A majority of the problems in thermodynamics are concerned with work that results from an expansion or compression of a fluid. For an illustration, consider a gas contained in the cylinder of Fig. 6.2. The gas is under a pressure P and it exerts a force F on area A of the piston face equal to PA. If the piston moves in increments dx from position 1 to position 2, the total work of the process can be found by applying Eq. (6.10):

$$ {}_1W_2 = \int_1^2 F \cdot dx = \int_1^2 PA\ dx $$

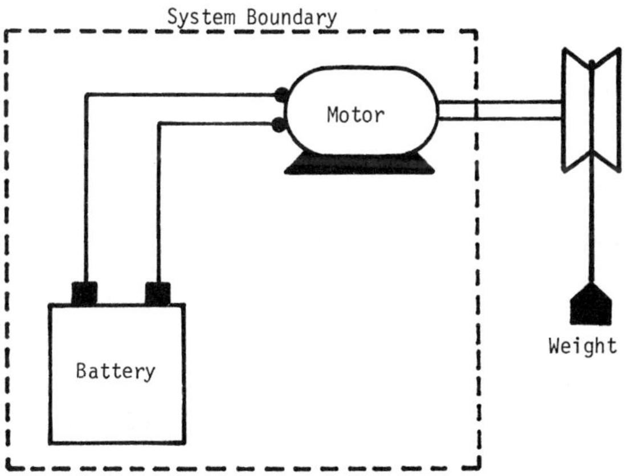

FIGURE 6.1 Battery-driven motor that raises or lowers a weight; illustrates chemical, electrical, and mechanical work.

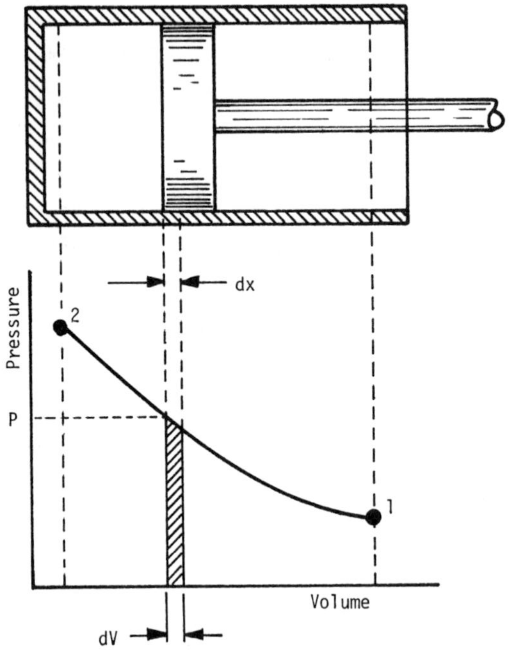

FIGURE 6.2 Work done by the piston in compressing the gas from state 1 to state 2 is the area under the curve.

However, A dx represents an infinitesimal change in the volume of the fluid,

$$\therefore \quad {}_1W_2 = \int_1^2 P\,dV \qquad (6.11)$$

When the functional relationship between the pressure and volume are known, the PV diagram of Fig. 6.2 can be drawn. Curve 1-2 is the path of the process, and the area under the curve is the work done by the fluid.

It is possible for the process to go from state 1 to state 2 along many different paths, and each one would trace a different area. Consequently, the amount of work done depends on the path that is followed rather than on the end points. For this reason work is called a *path function*. The term path function has a distinctly different meaning than the term *point function*, which applies to thermodynamic properties. For example, the total difference in volume between two states depends only on the initial and final volume, and any intermediate volumes are irrelevent:

$$\int_1^2 = dV = V_2 - V_1 \tag{6.12}$$

In mathematical terms dV is said to be an exact differential and its integral is equal to the difference in the end point values. The same relationship is not true for a path function such as work. Consequently, work is referred to as an inexact differential. To distinguish between the two forms, the symbol δW is used to designate the differential form of work. It follows that the total work associated with a process is

$$\int_1^2 \delta W = {}_1W_2 \tag{6.13}$$

Note that the expression $W_2 - W_1$ is not used to represent total work in thermal analysis.

Several ways in which work can be done were listed above. Mathematical equations expressing each of the different kinds of work have different forms because of the variables that are used for each. There are, however, several characteristics that the equations have in common:

Each expresses the work of a quasistatic process.

Each is the integral of the dot product of an intensive property and the change of an extensive property.

Work in each case is the area under a curve whose ordinate represents the intensive property and whose abcissa represents the extensive property.

Since work is a transfer of energy, the units for work are the same as those for energy. The time rate of work is called *power*, which is assigned the same units as time rates of energy such as watts and horsepower.

6.4 CONCEPT OF HEAT

Heat Q is defined as energy transferred across a boundary due to a temperature gradient between a system and its environment. Therefore, heat is energy in transit, and the modes of heat transfer are conduction, convection, radiation, and change of phase; these modes will be considered in later chapters.

In the previous section it was stated that work is not a function of system coordinates but depends on the path of the process between initial and final states. The same statement is true for heat. Consequently, Q is not a function of thermodynamic coordinates, but depends on the path. As in the case of work, an infinitesimal amount of heat is an inexact differential designated as δQ. The total heat $_1Q_2$ transferred during a process from state 1 to state 2 is expressed as

$$\int_1^2 \delta Q = {}_1Q_2 \tag{6.14}$$

Units for Q are the same as those for energy because heat is a form of energy transfer. In much of the calculation done in thermal technology, the heat transfer rate, rather than the amount of heat, is the important consideration. For heat transfer rate the appropriate units are cal/s, Btu/s, or watts (joule/s). A commonly accepted convention is to designate heat or heat transfer rate as positive when it is transferred to a system.

6.5 COMPARISON OF WORK AND HEAT

Heat and work are extremely important concepts in the study of thermal systems since they represent the sole means for transferring energy from or to a closed system. Although they are distinctly different phenomena, they have certain characteristics in common:

Heat and work are not thermodynamic properties; they are not conserved.

Heat and work are both transient phenomena. Systems cannot be said to contain heat or work, but either or both may occur when a system undergoes a change of state.

Heat and work are both boundary phenomena; they represent a change in energy across the boundary of a system.

Heat and work are methods by which the internal energy of a system is changed, but it is not possible to separate the internal energy into a heat or work portion.

Heat and work are the only two mechanisms for transferring energy from or to a closed system.

Heat and work are both path functions and, consequently, are expressed mathematically as inexact differentials δQ and δW.

Heat and work have opposite sign conventions. Heat is positive when it is transferred to a system and work is positive when it is transferred from a system.

The units for heat and work are the same as those for energy.

Heat and work can be related numerically by the mechanical equivalent of heat J:

$$J = 4.186 \text{ joule/cal} = 778.26 \text{ ft lbf/Btu} \tag{6.9}$$

6.6 HEAT CAPACITY AND SPECIFIC HEAT

Words such as heat capacity and specific heat are actually misnomers since it is meaningless to refer to the amount of heat in a substance. The ideas behind the words, however, are among the most important concepts in thermal technology [8,9]. When heat is transferred across the boundary of a system, the temperature of the system may or may not change depending on the nature of the process. For example, if heat is transferred to a saturated liquid, a change in state will occur, but the temperature will remain constant. In general, however, a transfer of heat from or to a system will result in a change of temperature.

If the temperature of a system changes from T_1 to T_2 when an amount of heat Q is transferred to the system, the *average heat capacity* is defined as

$$\text{Average heat capacity} \equiv \frac{Q}{T_2 - T_1} \tag{6.15}$$

By allowing Q and $\Delta T = T_2 - T_1$ to become very small, the *instantaneous heat capacity* C can be expressed mathematically as

$$C = \lim_{T_2 \to T_1} \frac{Q}{T_2 - T_1}$$

or

$$C = \frac{\delta Q}{dT} \tag{6.16}$$

In keeping with the usual practice in thermodynamics, a specific heat capacity (commonly called *specific heat* c) is defined as the heat capacity per unit mass:

$$c \equiv \frac{C}{m} = \frac{\delta Q}{m\ dT} \tag{6.17}$$

which, when integrated gives

$${}_1Q_2 = m \int_{T_1}^{T_2} c\ dT \tag{6.18}$$

At ordinary temperatures and over ordinary temperature intervals, heat capacity and specific heat can be considered constant so that finite relationships can be used:

$${}_1Q_2 = C\ \Delta T$$

$${}_1Q_2 = mc\ \Delta T \tag{6.19}$$

The value of heat capacity for a substance depends on the nature of the heat transfer process. Values of C, for example, will differ depending on whether the pressure of a substance is changed or kept constant.

A wide variety of heat capacities has been investigated for many different systems. For a stretched wire system, heat capacities at constant length and constant tension are important to researchers; for film systems, constant area and constant surface tension are important; for electrical systems, constant voltage and constant charge are important; etc.

Specific heats at constant pressure (c_P) and constant volume (c_V) have been found to be most useful for thermodynamic systems.

These quantities are defined in terms of changes in enthalpy and internal energy, respectively, as

$$c_p \equiv \left(\frac{\partial h}{\partial T}\right)_p \tag{6.20}$$

$$c_v \equiv \left(\frac{\partial u}{\partial T}\right)_v \tag{6.21}$$

and, once again, finite differences can be used:

$$\Delta h \simeq c_p \; \Delta T \tag{6.22}$$

$$\Delta u \simeq c_v \; \Delta T \tag{6.23}$$

Constant pressure and constant volume specific heats are thermodynamic properties since they are defined in terms of properties. Equations (6.22) and (6.23) neglect any electrical, magnetic, or surface effects. The field of measurement in which heat capacities and specific heats are obtained is called *calorimetry* and is one of the most exacting branches of experimental technology. In practice, values of c_p are measured because it is very difficult to measure c_v, particularly for solids. Values of c_v are obtained from relationships between c_p and c_v; these relationships will be covered in Chap. 9.

The definition of the calorie and the Btu provides the standard unit for specific heat. That is, 1.0 cal/gm °C or 1.0 Btu/lbm °F is the specific heat for water at 15°C; the value varies less than 1% between 0 and 100°C. Tables 6.1 and 6.2 show typical values of specific heat for fluids and solids; ref. 10 and 11 contain additional data.

6.7 HEATS OF TRANSFORMATION

Of special interest in engineering systems are the amounts of heat required to effect phase changes in various substances. These amounts are called *heats of transformation* and are very important in the fields of power generation, refrigeration, and boiling heat transfer. Special designations for heats of transformation are commonly used:

TABLE 6.1. Typical Values of Specific Heat for Engineering Substances at Ordinary Temperatures

Substance and Phase		c_v(Btu/lbm °F) or (cal/g °C)	c_p(Btu/lbm °F) or (cal/g °C)
Water	(ℓ)	1.0	1.0
Water	(g)	0.34	0.45
Gasoline	(ℓ)	-	0.5
Acetone	(ℓ)	-	0.54
Air	(g)	0.17	0.24
Freon 12	(ℓ)	0.22	0.22
CH_4	(g)	0.4	0.53
CO_2	(g)	0.16	0.2
Light Oil	(ℓ)	0.5	0.5
Machine Oil	(ℓ)	-	0.4

Conversion of Units

[cal/g °C] x (4.19) = [joule/g °C]

TABLE 6.2 Typical Values of Specific Heat for Engineering Solids

Solid	Specific Heat, (Btu/lbm °F) or (cal/g °C)	Temperature, °C
Ice	0.55	-10 to 0
Ice	0.45	-30
Aluminum	0.217	15 to 100
Glass	0.12	10 to 100
Copper	0.093	10 to 100
Lead	0.03	20 to 100
Brass	0.094	15 to 100
Silver	0.056	0 to 100

Conversion of Units

[cal/g °C] x (4.19) = [joule/g °C]

Heat of fusion is the heat absorbed to change a solid to a liquid.

Heat of vaporization is the heat absorbed to change a liquid to a gas.

Heat of sublimation is the heat absorbed to change a solid to a gas.

Heat of condensation is the heat liberated to change a gas to a liquid.

Heat of solidification is the heat liberated to change a liquid to a solid.

Heat of sublimation is the heat liberated to change a gas to a solid.

A great deal of experimental evidence shows that the amount of heat Q required to change the phase of a substance is proportional to the mass (m) of the substance. The heat of transformation L represents the proportionality constant needed to express the relationship numerically:

$$Q = mL \tag{6.24}$$

Table 6.3 lists values of L for several substances used in engineering systems.

When a substance changes phase, the temperature does not change. Consequently, the heat of transformation is sometimes referred to as *latent heat*. For example, the change in enthalpy at constant pressure h_{fg} is known as the latent *heat of vaporization*. By referring to tables of thermodynamic properties [1-3], it can be seen that h_{fg} is equal to the difference between the enthalpy of saturated vapor and saturated liquid ($h_{fg} = h_g - h_f$). This is a general relationship and the heat of transformation for any phase change equals the difference between the enthalpies of the phases:

$$L = h_2 - h_1 \tag{6.25}$$

For proof of Eq. (6.25) see solved problem 6 in Chap. 7.

A particularly simple relationship exists between the heats of transformation at the triple point of a substance. That is, the heat

TABLE 6.3 Approximate Values of Heats of Transformation for Various Substances at Atmospheric Pressure

Substance	Heat of Fusion (cal/g)	Heat of Vaporization (cal/g)
Aluminum	94	2261
Lead	5.9	1750
Teflon	417	-
Sodium	27	1006
Water	79	539
Hydrogen	14	108
Oxygen	3.3	51
Ethyl Alcohol	25	204

Conversion of Units

[cal/g] x (1.8) = [Btu/lbm]

[cal/g] x (4.19) = [joule/g]

of sublimation equals the sum of the heat of fusion and the heat of vaporization:

$$L_{13} = L_{12} + L_{23} \tag{6.26}$$

Where L_{13}, L_{12}, and L_{23} represent the heats of sublimation, fusion, and vaporization, respectively.

6.8 CONCLUSION

In this chapter the concepts of work and heat are introduced. Rather than being properties of a system, they are shown to be manifestations of changes in energy. Although they are distinctly different phenomena, they have several characteristics in common; these characteristics are outlined. Units and dimensions in several systems are considered and mathematical relations for work and heat are presented.

The net effect of applying work and heat to an engineering system can be expressed in terms of property changes. In addition to previously defined properties, new ones such as internal energy, flow energy, enthalpy, heat capacity, specific heat, and heat of transformation are introduced.

SOLVED PROBLEMS

PROBLEM 1: A body of mass 5 kg falls freely from a height of 10 m. Calculate the kinetic energy of the body as it strikes the ground, and show that it is equal to the initial potential energy.

Solution: The velocity of the body can be found from an equation of mechanics:

$$V^2 = 2gZ = 2(9.8\ m/s^2)(10\ m) = 196\ m^2/s^2$$

$$KE = \frac{mV^2}{2g_c} = \frac{(5\ kg)(196\ m^2/s^2)}{(2)(1.0\ kg\ m/N\ s^2)} = 490\ J$$

$$PE = \frac{mgZ}{g_c} = \frac{(5\ kg)(9.8\ m/s^2)(10\ m)}{1.0\ kg\ m/N\ s^2} = 490\ J$$

PROBLEM 2: At one state in a flow process steam at 149°F has a quality of 20%. What is the enthalpy?

Solution: Use Keenan and Keyes *Steam Tables* [1]:

$$h = h_f + h_{fg}$$

$$h = 116.96 + (0.20)(1008.7) = 318.7\ Btu/lbm$$

PROBLEM 3: A horizontal force of 30 lb moves a block weighing 40 lb a distance of 10 ft to the left in 2 s. Determine (a) the total work done, (b) the work done against friction, (c) the work done against gravity, and (d) the frictional horsepower. Assume the kinetic coefficient of friction is 0.1.

Solution: From mechanics $f = \mu N$

where

f = frictional force

μ = kinetic coefficient of friction = 0.1

N = normal force on the block = 40 lbf

∴ f = (0.1)(40) = 4 lbf

(a) Total work done = (30 lbf)(10 ft) - (4 lbf)(10 ft) = 260 ft lbf

(b) Frictional work = (4 lbf)(10 ft) = 40 ft lbf

(c) No work is done against gravity because the block has horizontal motion only.

(d) Frictional horsepower $= \frac{40 \text{ ft lbf}}{2 \text{ s}} = 20$ ft lbf/s

$$\left| 20 \ \frac{\text{ft lbf}}{\text{s}} \right| \left| \frac{\text{horsepower}}{550 \text{ ft lbf/s}} \right| = 0.036 \text{ hp}$$

PROBLEM 4: If all the frictional work in the previous problem is converted into heat, how many Btu's are involved?

Solution: The number of Btu's can be found by applying the mechanical equivalent of heat (J = 778 ft lbf/Btu):

$$\frac{40 \text{ ft lbf}}{778 \text{ ft lbf/Btu}} = 0.051 \text{ Btu}$$

PROBLEM 5: A 50 gm sample of metal is finely divided, heated at 98°C, and then dropped into 75 gm water at 19°C in a calorimeter. The final temperature of the mixture is 27°C. The mass of the calorimeter is 123 gm, and its specific heat is 0.1 cal gm^{-1} $°C^{-1}$. If no heat is lost, determine the mean specific heat of the metal sample.

Solution: The heat lost by the metal sample is absorbed by the calorimeter, the water, and the instrumentation:

$$Q_{metal} = Q_{cal} + Q_{water} + Q_{inst}$$

$$Q = mc \ \Delta T$$

$$(50 \ c)(98 - 27) = [(123)(0.1) + (75)(1.0) + (6.5)]$$

$$\therefore \quad c = 0.21 \text{ cal/°C gm}$$

Note that only magnitudes have been considered. According to the usual convention, heat loss is negative and heat absorbed is positive.

PROBLEM 6: A beaker contains 500 gm of H_2O at 80°C. How many grams of ice at -4°F must be added to the water so that the final temperature will be 50°C?

Solution: Neglect the specific heat of the beaker and assume that all the heat lost by the ice is absorbed by the water. Convert the ice temperature to °C:

$$\frac{5}{9}(-4 - 32) = -20°C$$

The ice will melt and so the heat of transformation must be applied to account for the liberation of heat during the phase change:

$$Q = mc\ \Delta T + mL \quad \text{and} \quad Q_{ice} = Q_{H_2O}$$

$$(mc\ \Delta T)_{ice} + (mL)_{ice} = (mc\ \Delta T)_{H_2O}$$

where for ice

$$c = 0.5 \text{ cal/gm °C @ } -20°C$$

$$\Delta T = -20 - 50°C = -70°C$$

$$L = 79.7 \text{ cal/gm}$$

where for H_2O

$$m = 500 \text{ gm}$$

$$c = 1.0 \text{ cal/gm}$$

$$\Delta T = 80 - 50°C = 30°C$$

Using magnitudes:

$$(m)(0.5)(70) + (m)(79.7) = (500)(1.0)(30)$$

$$\therefore \quad m = 131 \text{ gm}$$

EXERCISES

1. What velocity must be attained by a 1 lb mass if it is to have a kinetic energy of one Btu?
2. To what elevation must a mass of 1 kg be raised if it is to have a potential energy of 1 J?
3. A body whose mass is 5 slugs falls freely from a height of 20 ft. Calculate the kinetic energy at the end of the fall and show that it equals the initial potential energy.
4. Determine the internal energy of a quantity of steam whose pressure is 75 psi and whose quality is 85%.

5. What is the flow energy of 1 lbm steam in a rigid tank?
6. What is the mechanical equivalent of heat?
7. Steam in a flow process has a temperature of 125°C and a quality of 85%. What is the enthalpy?
8. Knowing that 1 kWh equals 860 kcal, and that the average human dissipates 2500 kcal per day, show that the average human gives off energy at the same rate as a 125-W light bulb.
9. A force of 50N, inclined 20° to the horizontal, moves a 50 kg mass a distance of 2 m along a horizontal surface in 4 s. What is the work done and the power dissipated?
10. How much heat is contained in a 1-lb mass of copper at 70°F?

REFERENCES

1. J. H. Keenan, F. G. Keyes, P. G. Hill, and J. C. Moore, *Steam Tables*, Wiley, New York, 1969.
2. J. H. Keenan and J. Kaye, *Gas Tables*, Wiley, New York, 1948.
3. *Refrigerant Tables, Charts, and Characteristics*, ASHRAE, 1959.
4. G. J. VanWylen, *Thermodynamics*, pp. 71-82, Wiley, New York, 1959.
5. *Handbook of Chemistry and Physics*, pp. F-214 to F-236, Chemical Rubber Co., Ohio, 1969.
6. W. M. Rohsenow and J. P. Hartnett, *Handbook of Heat Transfer*, pp. 1-15 to 1-17 and 2-3 to 2-7, McGraw-Hill, New York, 1973.
7. *Marks Standard Handbook for Mechanical Engineers*, pp. 1-73 to 1-85, McGraw-Hill, New York, 1969.
8. M. W. Zemansky, *Heat and Thermodynamics*, pp. 70-76, McGraw-Hill, New York, 1957.
9. F. W. Sears, *Thermodynamics*, 2d ed., pp. 46-52, Addison-Wesley, Reading, Mass., 1964.
10. *Handbook of Chemistry and Physics*, pp. D-94 to D-101, Chemical Rubber Co., Ohio, 1969.
11. *Marks Standard Handbook for Mechanical Engineers*, pp. 4-11, 4-12, 6-10, 6-67, 6-209, and 6-215, McGraw-Hill, New York, 1969.

7

FIRST LAW OF THERMODYNAMICS

7.1 INTRODUCTION

The first law of thermodynamics states that the change in the energy of a system exactly equals the net flow of energy across the boundary of the system; in effect, it says that energy is always conserved. The validity of the first law is based on experimental evidence, as are all scientific laws. Results of many experiments agree with the statement of the law, and no results have been found to contradict it.

7.2 MATHEMATICAL FORMULATION

A qualitative statement of a law is not sufficient to provide quantitative results. Any meaningful analysis requires a mathematical formulation. We can obtain a formulation by considering a closed system and applying some of the concepts covered in previous chapters.

In our hypothetical system there is a change in state from point 1 to point 2 by means of process a. A return from state 2 to state 1 can take place by either process b or c. These processes are shown in Fig. 7.1, where an intensive thermodynamic property is plotted against an extensive thermodynamic property. Typical coordinates can be pressure vs volume, surface tension vs film area, potential

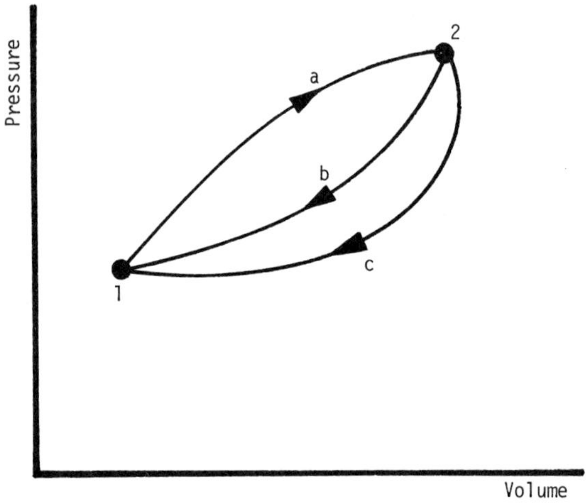

FIGURE 7.1 An intensive property vs an extensive property showing two of an infinite number of possible cycles. Diagram used to show the existence of the thermodynamic property, total energy.

difference vs electric charge, etc. Work is represented as the area under any of these curves.

In Fig. 7.1 a cycle takes place when either processes a and b or processes a and c are combined. Suppose, for example, that the system is a tank of water in which a paddle can be turned by means of an external falling weight. As work is done by the falling weight, heat is transferred to the water, raising its temperature. Such a system was used by Joule to demonstrate that work and heat are different manifestations of energy. When heat is transferred from the system, the cycle is completed and the net area under the two process curves represents the energy stored in the system. An infinite number of experiments of this type can be visualized, and cycles a-b and a-c of Fig. 7.1 represent just two of them.

Since heat and work are the only means of changing the energy of a closed system, the first law implies that the heat associated with a particular cycle must equal the work associated with the same cycle:

$$({}_1Q_2)_a + ({}_2Q_1)_b = ({}_1W_2)_a + ({}_2W_1)_b$$

and

$$({}_1Q_2)_a + ({}_2Q_1)_c = ({}_1W_2)_a + ({}_2W_1)_b$$

The second of these equations can be subtracted from the first:

$$({}_2Q_1)_b - ({}_2Q_1)_c = ({}_2W_1)_b - ({}_2W_1)_c$$

which can be rearranged to

$$({}_2Q_1)_b - ({}_2W_1)_b = ({}_2Q_1)_c - ({}_2W_1)_c$$

or

$${}_2Q_1 - {}_2W_1 = \text{constant}$$

It also follows from Eqs. (6.13) and (6.14) that $\delta Q - \delta W = \text{constant}$.

The relation above states that the difference between the total heat and total work for states 1 and 2 in a closed system does not depend on the process between the states. The difference satisfies the characteristics of a thermodynamic property, and we identify it as total energy change $(E_2 - E_1)$. The difference can also be expressed in terms of an exact differential dE. The basic mathematical formulation of the first law of thermodynamics is written as

$$\delta Q = dE + \delta W \tag{7.1}$$

$${}_1Q_2 = (E_2 - E_1) + {}_1W_2 \tag{7.2}$$

7.3 FIRST LAW FOR AN OPEN SYSTEM

The significance of property E from the previous section is that it represents the total energy of a system, including mechanical, electrical, chemical, or any other type of energy. Equations (7.1) and (7.2) can be extended to an open system if an accounting is made of the energy that enters and leaves the system as a result of the passage of matter across the system boundaries.

Practitioners of various disciplines are interested in particular forms of energy. For example, a turbine designer is concerned with kinetic and potential energies, a physical chemist with surface and strain energies, an electronic circuit designer with electrical and magnetic energies, and so forth. Since this book is concerned with thermal technology and its application, the conventional practice of considering total energy E to be made up of internal, flow, kinetic, and potential energies will be followed:

$$E = U + PV + KE + PE \tag{7.3}$$

Specific internal energy u for a variety of substances can be found in various publications [1-5], and kinetic and potential energies can be calculated from Eqs. (6.1) and (6.2). Flow energy PV is the product of pressure and volume, although in almost all instances the property of enthalpy is substituted for the sum of internal and flow energies, as expressed by Eqs. (6.6) and (6.7). Keep in mind that Eq. (7.3) expresses the total energy of a system in a form that is convenient for thermodynamicists and thermal technologists. It is possible to apportion E in many different ways.

Figure 7.2 is a schematic illustration of an open system, showing all the ways that the energy of the system can be changed. Net heat δQ supplied to the system and net work δW done by the system are considered as positive quantities. Subscripts i and o refer to conditions at the inlet and outlet of the system, respectively, and subscript s refers to conditions within the system.

When a mass of substance δm_i enters the system, there is an increase in energy by the amount

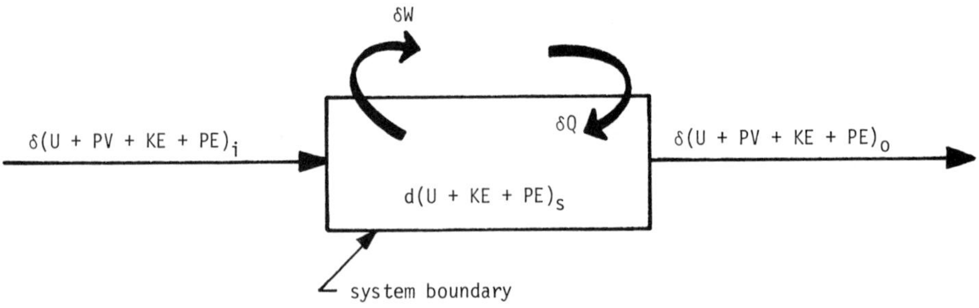

FIGURE 7.2 Schematic illustration of an open system.

$$\delta(U + PV + KE + PE)_i$$

and a decrease in energy when a mass m_o leaves the system of

$$\delta(U + PV + KE + PE)_o$$

Within the system a mass of substance dm_s undergoes an increase or decrease in energy of the amount

$$d(U + KE + PE)_s$$

Since the mass within the system does not cross a system boundary, there is no flow-energy term included. Note that once again we must distinguish between the infinitesimal operators δ and d. The mass entering and leaving is taken as δm_i and δm_o, while the mass which undergoes an energy change within the system is dm_s. Operator d refers to a change in a property, while operator δ signifies a transfer of some differential amount. The clearest way to see the difference between the operators is to see how they are integrated. The integral of d represents the total change in a property and can be expressed by the difference operator Δ; the integral of δ is simply the total amount of whatever enters or leaves the system.

The general differential form of the first law of thermodynamics for an open system can be written by applying the principle of conservation of energy to Fig. 7.2:

$$\delta Q + \delta(U + PV + KE + PE)_i = \delta W + \delta(U + PV + KE + PE)_o + d(U + KE + PE)_s \tag{7.4}$$

Equation (7.4) is a general equation and applies equally well to open and closed systems. For the latter case, δm_i and δm_o are zero. Despite its generality, Eq. (7.4) does not have a convenient form, and so a modified version is customarily used for thermal analyses. First of all, it is convenient to identify the mass that enters and leaves the system. This can be done by factoring the mass from each of the total energy terms. Also, the property of specific enthalpy h replaces the sum of internal and flow energies according to Eqs. (6.6) and (6.7). There is no enthalpy associated with the mass dm_s because it does not cross a boundary of the system. However, internal energy U can be expressed as the product of mass and specific internal energy.

With the above revisions, the differential form of the first law for an open system appears as

$$\delta Q + \delta m_i \left(h + \frac{V^2}{2g_c} + \frac{gZ}{g_c}\right)_i = \delta W + \delta m_o \left(h + \frac{V^2}{2g_c} + \frac{gZ}{g_c}\right)_o + d(mu + KE + PE)_s \tag{7.5}$$

and the integrated form appears as

$$ {}_1Q_2 + m_i \left(h + \frac{V^2}{2g_c} + \frac{gZ}{g_c}\right)_i = {}_1W_2 + m_o \left(h + \frac{V^2}{2g_c} + \frac{gZ}{g_c}\right)_o + (U_2 - U_1 + KE_2 - KE_1 + PE_2 - PE_1)_s \tag{7.6}$$

For many engineering applications the first law has to be considered as a rate equation. A rate form can be developed by considering a time period $d\tau$ during which all of the energy changes occur. When $d\tau$ is very small, the following rates can be defined:

$$\text{Heat transfer rate} = \frac{\delta Q}{d\tau} = \dot{Q}$$

$$\text{Work transfer rate} = \frac{\delta W}{d\tau} = \dot{W}$$

$$\text{Mass transfer rate} = \frac{\delta m}{d\tau} = \dot{m}$$

The rate form of the first law for an open system results when the transfer rates shown above are placed in Eq. (7.5):

$$\dot{Q} + \dot{m}_i \left(h + \frac{V^2}{2g_c} + \frac{gZ}{g_c}\right)_i = \dot{W} + \dot{m}_o \left(h + \frac{V^2}{2g_c} + \frac{gZ}{g_c}\right)_o + \frac{d}{d\tau}(mu + KE + PE)_s \tag{7.7}$$

7.4 SPECIAL FORMS OF THE FIRST LAW

Three forms of the first law of thermodynamics developed in the previous section can be applied to any kind of a system. In practice, however, it is rarely necessary to use any of these equations in their entirety. Some simplification such as no mass flow, steady state condition, negligible velocity or elevation can be made so that a less complex form of the first law equation is sufficient.

Closed Systems

When there is no mass flow into or out of a system, a closed system results. Differential and integrated forms of the first law for such systems can be developed by letting δm_i and δm_o equal zero in Eqs. (7.5) and (7.6):

$$\delta Q = \delta W + d(mu + KE + PE)_s \tag{7.8}$$

$$_1Q_2 = {_1W_2} + (\Delta U + \Delta KE + \Delta PE)_s \tag{7.9}$$

Steady Flow Systems

Many devices are encountered in engineering practice through which there is a steady flow of fluid. Typical examples are turbines, pumps, fans, nozzles, and orifices. A process can be considered as one of steady flow when the following conditions are satisfied:

Mass transfer rate is constant at all sections of the device ($\dot{m}_i = \dot{m}_o = \dot{m}$).

Properties of the fluid at any section remain constant, but may differ from one section to another.

Total energy of the system remains unchanged; $(dE/d\tau)_s = 0$.

Energy transfer rates are constant [$\dot{Q}$ and $\dot{W} \neq f(\tau)$].

If the above conditions are applied to Eq. (7.8), the result is the first law for a steady flow process:

$$\dot{Q} + \dot{m}\left(h + \frac{V^2}{2g_c} + \frac{gZ}{g_c}\right)_i = \dot{W} + \dot{m}\left(h + \frac{V^2}{2g_c} + \frac{gZ}{g_c}\right)_o \tag{7.10}$$

Miscellaneous Systems

In addition to the special cases mentioned above, there are other conditions that simplify the first law equations. For example, in many devices the change in elevation is small, and so the potential energy change can be neglected. The velocity in some devices is low, or the change in velocity through the system is small; consequently, the kinetic energy change is insignificant. Finally, there are systems in which there is a negligible transfer of heat or work.

Certain characteristics of a process can help to solve first law problems. If, for example, temperature, pressure, volume, or other properties are constant, the calculations will be simplified. Obviously, one must understand the processes that take place and be able to account for all relevant energy transfers.

7.5 ENERGY BALANCE ANALYSIS

An extremely important procedure in engineering technology is the compilation of an energy balance for a system. The first law of thermodynamics, in its various forms, makes such a compilation possible. In the preceding sections a number of equations [(7.1) to (7.10)] have been developed which account for energy transfer in a variety of systems. The first two of these equations [(7.1) and (7.2)] can be applied to all types of systems with all types of energy transfer. The remainder of the equations are tailored to handle fluid flows in thermodynamic systems.

Two main types of systems are identified: closed and open systems. In some of the literature [6-8] these systems are referred to as control mass and control volume systems.

When an energy balance analysis is to be made, the first step is to identify the system, its boundary, and its environment. In most cases it is advisable to represent these elements by means of a sketch. A dotted line indicates the boundary that separates the system from its environment. Energy transfer is shown by appropriately drawn arrows to show the direction. All relevant energy transformations must be accounted for within the system and across its boundary.

These transformations can be shown by mathematical relationships or by process curves and diagrams.

The type of units to be used is an important consideration when energy transformations are being handled. There are characteristic units for various kinds of processes. For example, volts, amperes, and coulombs are customarily used with electrical systems, whereas newtons and meters, or pounds of force and feet, are used to describe mechanical systems. In addition, thermodynamic properties are available in joules per gram or Btu per pound mass, whereas work, kinetic energy, and potential energy are generally given in joules, ergs, and foot pounds of force. Some authors include the mechanical equivalent of heat (J factor) in equations for the first law. Although that practice has not been followed in this book, it is imperative that a homogeneous system of units be used in all calculations.

An essential part of an energy balance is the time period during which the energy is transferred. It is possible to compile the balance for a particular finite period or calculations can be based on instantaneous considerations.

After all the above considerations are made, an appropriate form of the first law can be selected from Eqs. (7.1) to (7.10). Certain idealizations and simplifications can then be made (e.g., negligible change in potential energy) and the equation solved.

In summary, steps taken in an energy balance of a system are listed below:

Identify the system, its boundary, and its environment.
Identify all relevant energy transformations in the system and the environment.
Make a sketch of the system and show all energy transfers in the system and across the boundary.
Decide on the system of units to be used.
Identify the period of time for the energy transfer.
Select a convenient form of the first law [Eqs. (7.1) to (7.10)].
Simplify the first law equation by making any appropriate idealizations.
Solve the problem and check the solution.

7.6 CONCLUSION

In this chapter qualitative and quantitative statements of the first law of thermodynamics are made. Initially, mathematical formulations of the law for an open system are developed, followed by a presentation of special forms of the first law for closed, steady state, and miscellaneous systems.

Use of the first law in compiling an energy balance for various kinds of systems is described, and it is shown to be an accounting system for the energy transfer from, to, and within a system. Detailed steps for making an energy balance are listed.

SOLVED PROBLEMS

PROBLEM 1: An automobile with a mass of 3000 lb comes over the top of a hill 50 ft high with a velocity of 50 mph. Brakes are applied at the instant the automobile reaches the top, and it comes to rest at the bottom of the hill. How much energy is dissipated from the brakes?

Solution: Consider the automobile a closed system and apply the first law [Eq. (7.4)]:

$$ {}_1Q_2 = U_2 - U_1 + m\left(\frac{V_2^2 - V_1^2}{2g_c}\right) + \frac{mg}{g_c}(Z_2 - Z_1) + {}_1W_2 $$

where

$m = 3000$ lbm

$V_1 = 50$ mph $= 73.3$ fps

$V_2 = 0$

$g = 32.2\ \mathrm{ft/s^2}$

$g_c = 32.2\ \mathrm{lbm\ ft/lbf\ s^2}$

$Z_1 = 50$ ft

$Z_2 = 0$

Neglecting windage and frictional effects, the change in internal energy and the work are zero; therefore, the heat transferred is a function of changes in kinetic and potential energies.

$$ {}_1Q_2 = \frac{3000}{(2)(32.2)} [0 - (73.3)^2] + \frac{(3000)(32.2)}{32.2} (0 - 50) $$

$$ = 4 \times 10^5 \text{ ft lbf} $$

PROBLEM 2: Steam at a pressure of 13 bar and a temperature of 300°C flows adiabatically and with negligible velocity into an evacuated tank. Using a closed system analysis, determine the temperature of the steam in the tank when the tank reaches line pressure.

Solution: Consider the mass of steam and the tank a closed system and apply the first law [Eq. (7.4)]:

$$ {}_1Q_2 = (U_2 - U_1) + \Delta KE + \Delta PE + {}_1W_2 $$

Changes in kinetic and potential energy can be neglected and ${}_1Q_2$ is zero for an adiabatic process. The steam in the line does work on the closed system by moving its boundaries. The work is negative and equal to $-p_1V_1$:

$$ 0 = U_2 - U_1 - p_1V_1 $$

which can be expressed in terms of specific quantities

$$ 0 = mu_2 - mu_1 - mp_1V_1 $$

but,

$$ h_1 \equiv u_1 + p_1V_1, $$

so

$$ u_2 = h_1 $$

That is, the final internal energy of the steam in the tank equals the initial enthalpy of the steam. From Keenan and Keyes *Steam Tables* [2]:

$$ h_1 \text{ at 13 bar and 300°C} = 3043.1 \text{ J/gm} $$

$$ \therefore \qquad u_2 = h_1 = 3043.1 \text{ J/gm} $$

From the steam tables (superheated vapor) at 13 bar pressure and an internal energy of 3043.1 J/gm, the temperature is 453.4°C (by interpolation).

PROBLEM 3: A 10 lb mass of air at 120°F is contained in a rigid tank. How much heat must be transferred to the tank to raise the air temperature to 275°F?

Solution: Let the air in the tank be a closed system and apply the first law [Eq. (7.9)]:

$${}_1Q_2 = {}_1W_2 + \Delta U + \Delta KE + \Delta PE$$

Since there is no change in volume, ${}_1W_2 = 0$, and there is no change in KE or PE:

$${}_1Q_2 = m(u_2 - u_1)$$

from Keenan and Kay *Gas Tables* [3]:

$$u_1 \text{ at } 120°F = 98.9 \text{ Btu/lbm}$$

$$u_2 \text{ at } 275°F = 125.6 \text{ Btu/lbm}$$

$${}_1Q_2 = (10)(125.6 - 98.9) = 267 \text{ Btu}$$

PROBLEM 4: Fluid enters a turbine with a velocity of 1.0 m/s and an enthalpy of 2000 J/gm; it leaves with a velocity of 60 m/s and enthalpy of 1800 J/gm. Heat losses are 500 J/s, and the flow rate is 0.5 kg/s. If the inlet of the turbine is 3 m higher than the outlet, what is the maximum theoretical power that can be developed?

Solution: Consider the turbine as a steady flow system and apply the first law [Eq. (7.10)].

$$\dot{W} = \dot{Q} + \dot{m}\left[(h_i - h_o) + \frac{V_i^2 - V_o^2}{2g_c} + \frac{g}{g_c}(Z_i - Z_o)\right]$$

where

$\dot{W}$ = power developed

$\dot{Q}$ = -500 J/s (heat loss)

$$\dot{m} = 0.5 \text{ kg/s} = 500 \text{ gm/s}$$

$$h_i - h_o = 2000 - 1800 = 200 \text{ J/gm}$$

$$v_i^2 - v_o^2 = 60^2 - 1^2 = 3599 \text{ m}^2/\text{s}^2$$

$$Z_i - Z_o = 3 - 0 = 3 \text{ m}$$

$$g = 9.8 \text{ m/s}^2$$

$$g_c = 10^3 \text{ gm m/Ns}^2$$

$$\dot{W} = -500 + 500\left[(200) + \frac{3599}{(2)(10^3)} + \frac{9.8}{10^3}(3)\right] = 1.0 \times 10^5 \text{ J/s}$$

PROBLEM 5: To produce 0.3 liter/s hot water at 82°C, low pressure steam at 2.4 bar and 80% quality is mixed with a stream of water at 16°C. What is the required steam flow rate?

Solution: Select an open system for analysis in which no work is done and no heat is lost; also, changes in potential and kinetic energy are neglected:

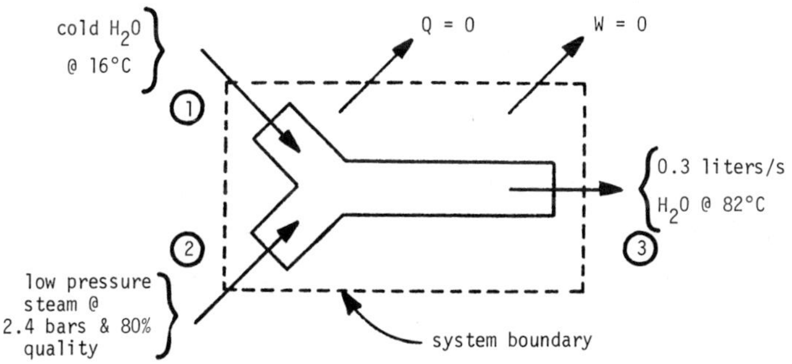

Since there is no change in stored energy, an abbreviated form of the first law [Eq. (7.7)] can be applied to the open system:

$$\dot{m}_1 h_1 = \dot{m}_2 h_2 = \dot{m}_3 h_3$$

where

$\dot{m}_1$ = water flow rate

h_1 = 67.19 J/gm (liquid water at 16°C)

$\dot{m}_2$ = steam flow rate

h_1 = 67.19 J/gm (liquid water at 16°C)

$\dot{m}_2$ = steam flow rate

h_2 = enthalpy of steam at 80% quality and 2.4 bar

$\dot{m}_3$ = mixture flow rate

h_3 = 343.3 J/gm (liquid water at 82°C)

h_f at 2.4 bars = 529.65 J/gm

h_{fg} at 2.4 bars = 2185.4 J/gm

v_3 for sat liquid at 82°C = 1.0305 cm^3/gm

$$h_2 = h_f + xh_{fg} = 529.65 + (0.8)(2185.4) = 2278 \text{ J/gm}$$

$$\dot{m}_3 = \frac{\dot{V}_3}{v_3} = \frac{300 \text{ cm}^3/\text{s}}{1.0305 \text{ cm}^3/\text{gm}} = 291.12 \text{ gm/s}$$

$\dot{m}_1 = \dot{m}_3 - \dot{m}_2$ (mass flow rate balance)

Substituting into the modified first law:

$$(291.12 - \dot{m}_2)(67.19) + \dot{m}_2(2278) = (291.12)(343.3)$$

$$\therefore \quad \dot{m}_2 = 36.4 \text{ gm/s}$$

PROBLEM 6: Latent heat of transformation can be defined as the ratio of heat absorbed to the mass which undergoes a change of phase (L = Q/m). Show that the heat of transformation for any phase change equals the difference between the enthalpies of the system in the two phases.

Solution: Phase changes occur at constant temperature and pressure, and so the specific work done can be expressed as

$$W/m = p(v_2 - v_1)$$

Neglecting any changes in KE or PE, the change in internal energy can be found by applying the first law:

$$u_2 - u_1 = \frac{Q}{m} - \frac{W}{m}$$

or

$$u_2 - u_1 = L - p(v_2 - v_1)$$

$$L = (u_2 + pv_2) - (u_1 + pv_1)$$

and from the definition of enthalpy

$$L = h_2 - h_1$$

EXERCISES

1. Define the total energy change for a closed system.
2. Give three examples of a closed system.
3. An automobile of 1500 kgm is traveling at 80 km/hr along a horizontal surface when the brakes are applied. Determine the amount of energy dissipated when the automobile comes to rest.
4. Show how the first law of thermodynamics for a closed system can be applied to a stream of water flowing over a 50-m embankment. What happens to the kinetic energy that is developed?
5. A rock on an embankment falls into a lake, and eventually the noise and splashing cease and equilibrium is attained. Show how the first law can be applied to the process.
6. Superheated steam at 300 psia and 450°F flows adiabatically into an evacuated vessel. Using closed system analysis, determine the temperature of the steam in the tank after it reaches line pressure.
7. Work at the rate of 2 kW is supplied to a fluid in a rigid tank while heat is being dissipated at the rate of 0.4 kW. What is the rate of change of internal energy?
8. A rigid tank contains 5 lbm air at 150°F. What is the final temperature of the air if 300 Btu of heat is transferred to the tank.
9. Fluid enters a turbine with a velocity of 6 fps and an enthalpy of 850 Btu/lbm; it leaves with a velocity of 280 fps and an enthalpy of 800 Btu/lbm. Heat is lost at the rate of 25 Btu/hr

and the flow rate is 2.0 lbm/s. If the inlet of the turbine is 10 ft above the outlet, what is the theoretical horsepower that can be developed?

10. To produce 5 gpm hot water at 180°F, low pressure steam at 35 psia and 80% quality is mixed with a stream of water at 60°F. Determine the required steam flow rate.

REFERENCES

1. J. H. Keenan, F. G. Keyes, P. G. Hill, and J. G. Moore, *Steam Tables - English Units*, Wiley, New York, 1969.
2. J. H. Keenan, F. G. Keyes, P. G. Hill, and J. G. Moore, *Steam Tables - Metric Units*, Wiley, New York, 1969.
3. J. H. Keenan and J. Kaye, *Gas Tables*, Wiley, New York, 1948.
4. *Refrigerant Tables, Charts, and Characteristics*, ASHRAE, 1959.
5. *Handbook of Physics and Chemistry*, pp. E12-E27, Chemical Rubber Co., Ohio, 1969.
6. G. J. Van Wylen, *Thermodynamics*, pp. 64-118 and 517-523, Wiley, New York, 1959.
7. W. C. Reynolds, *Thermodynamics*, pp. 77-115, McGraw-Hill, New York, 1965.
8. R. E. Balzhiser and M. R. Samuels, pp. 94-134, *Engineering Thermodynamics*, Prentice-Hall, Englewood Cliffs, N.J.

8

SECOND LAW OF THERMODYNAMICS

8.1 INTRODUCTION

The first law of thermodynamics is very useful in making quantitative evaluations of energy transfer. It makes no statement, however, about the direction in which the transfer can proceed. There is much experimental evidence to show that every process in nature has a preferred sense of direction. In this chapter we shall introduce the second law of thermodynamics, which predicts the directions in which energy transfers tend to proceed. The second law is not limited to qualitative predictions; mathematical formulations of the law can be used to predict the maximum efficiency of a process, the minimum work required to complete a process, or the final composition of a chemical reaction.

Before the second law can be discussed, several concepts associated with it must be defined. Among these are reversibility, heat engines and pumps, entropy, and an ideal cycle.

8.2 REVERSIBILITY AND IRREVERSIBILITY

When energy is transferred from a system, it is possible to do work. The amount of work done depends on the type of process that is used. The maximum amount of work that can be obtained from the process is called *available energy* or *availability*. Recently, some authors have used the name *exergy*.

Any real process will result in a loss of some of the available energy, and the loss is called the *irreversibility* of the process. System designers must give careful attention to keeping irreversibility at a minimum. It is not possible, however, to completely eliminate irreversibility since it is a natural occurrence in any real process. Figure 8.1 shows two examples of irreversible processes. In Fig. 8.1a electric current flows through a resistance due to the potential energy of a charged capacitor. All of this energy is eventually dissipated in the form of heat loss from the resistance. In Fig. 8.1b a block oscillates under the impetus of a spring. Amplitudes of the oscillation diminish uniformly until the block finally comes to rest. The initial potential energy in the spring is dissipated in the form of heat generated by the sliding friction of the block (and some friction in the spring). Obviously, it is not possible to completely eliminate the resistance in an electrical process or the friction in a mechanical process, and so a certain amount of irreversibility always exists. Numerous other processes have inherent irreversibilities: unrestrained expansion of a gas, heat transfer through a finite temperature difference, spontaneous chemical reaction, diffusion of one substance into another, etc.

All irreversible processes have one thing in common: the initial state of the system cannot be restored unless something is done

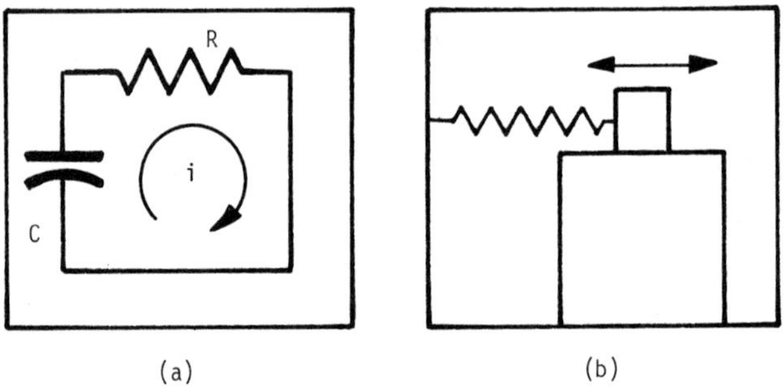

FIGURE 8.1 Irreversible processes: (a) current flow through a resistance, and (b) block with sliding friction.

that changes the environment of the system. For example, the capacitor of Fig. 8.1 could be recharged if a battery is connected to the circuit, or the spring-mass system could be restored to its initial position if work from outside the system is applied. These considerations of irreversibility allow us to define reversibility. A *reversible* process is defined as a process which can occur, and then be reversed, without any change to the system or its environment.

The only conceivable way to create a reversible process is to neglect the characteristics that cause irreversibility; two such idealizations are shown in Fig. 8.2. An electrical circuit with a charged capacitor, an inductor, and negligible resistance (Fig. 8.2a) represents a reversible oscillator. Electrical energy is continuously transferred between the capacitance and the inductance. In Fig. 8.2b a spring-mass system with negligible sliding friction also represents a reversible oscillator.

Idealizations can be made to remove other causes of irreversibility. Even though a reversible process can never be achieved in practice, it is an important concept since it represents a point of reference. It permits the development of an ideal process against which all other processes can be evaluated.

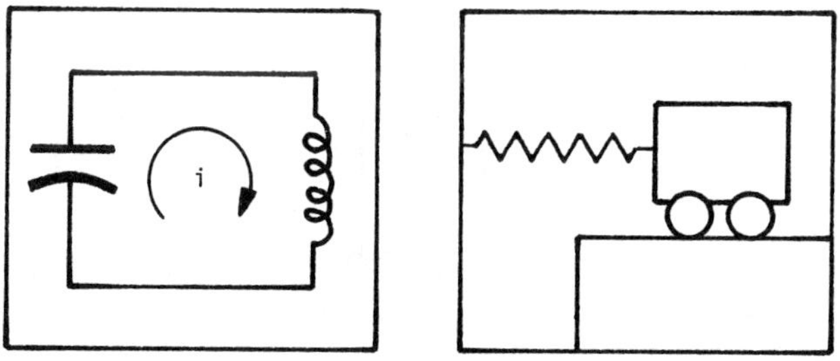

(a) (b)

FIGURE 8.2 Reversible processes: (a) electric curcuit with negligible resistance, and (b) block with negligible sliding resistance.

8.3 HEAT ENGINES AND HEAT PUMPS

The most useful thermodynamic devices are those that operate in cycles. A cycle has previously been defined as a system that passes through a succession of processes and returns to its initial state. Heat engines and heat pumps are two major types of devices considered in thermodynamics.

Van Wylen [1] succinctly defines a *heat engine* as a device that operates in a thermodynamic cycle and does a certain amount of work as a result of heat transfer from a high temperature body to a low temperature body. The bodies referred to in the definition can also be called thermal reservoirs. A *thermal reservoir* is a body from which, and to which, heat can be transferred for an indefinite period without any change in the body temperature. The term *heat source* is used for a reservoir from which heat is transferred and the term *heat sink* for a reservoir to which heat is transferred.

An example of a heat engine is a steam turbine, shown schematically in Fig. 8.3. The working substance is water, which receives heat

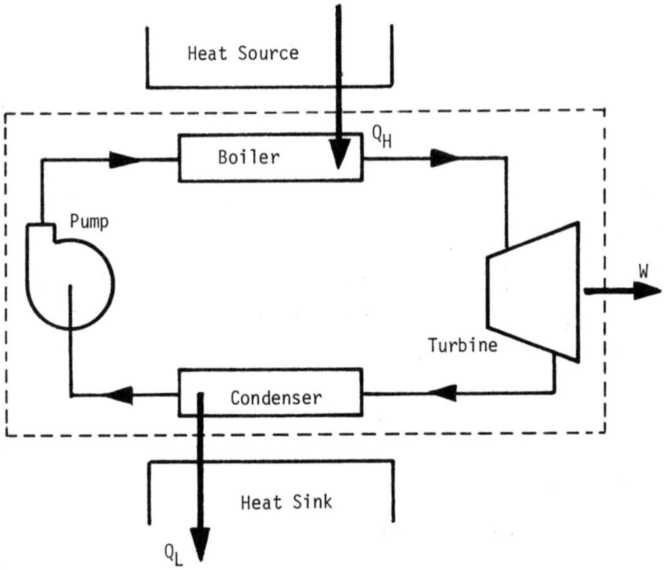

FIGURE 8.3 Schematic diagram of a heat engine.

from a heat source, thereby becoming steam. The steam passes through the turbine causing the turbine to deliver mechanical work. At this stage of the cycle some of the steam condenses. The resulting two-phase mixture proceeds to the condenser where the remaining vapor is liquefied. During this operation the heat of condensation is transferred to a heat sink. A pump then moves the liquid back into the boiler, completing the cycle. Figure 8.3 represents a closed system since only heat and work cross the boundary. Heat input is Q_H and heat rejected is Q_L; work W is a useful output from the turbine. Applying the first law in the form of Eq. (7.9), it is evident that, for a cycle, all quantities except work and heat are zero:

$$W = Q_H - Q_L$$

Thermal efficiency η_t for the system is defined in the usual terms of output over input:

$$\eta_t = \frac{W}{Q_h} = \frac{Q_H - Q_L}{Q_H} = 1 - \frac{Q_L}{Q_H} \qquad (8.1)$$

In all real systems friction losses cause the overall efficiency to be lower than the thermal efficiency.

Once again we refer to Van Wylen [1] for a succinct definition of a heat pump: a *heat pump* is a device that operates in a thermodynamic cycle but which requires input work and which accomplishes the objective of transferring heat from a low-temperature source to a high-temperature source. Essentially, a heat pump is a heat engine in reverse.

The working substance in a heat pump is called a *refrigerant*. A variety of substances are used commercially as refrigerants; a few of them are fluorinated hydrocarbons, ammonia, sulfur dioxide, and carbon dioxide.

A schematic illustration of a heat-pump cycle is shown in Fig. 8.4. Liquid refrigerant from the condenser is throttled in an expansion valve to a low pressure. The value of the pressure is determined by the desired boiling temperature of the refrigerant in the evaporator.

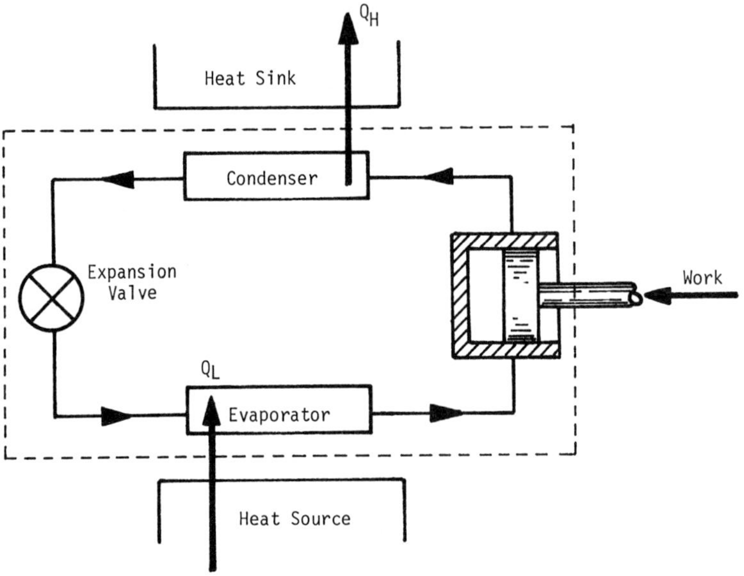

FIGURE 8.4 Schematic diagram of a heat pump.

Useful heat pumping is accomplished in the evaporator where heat is absorbed from the low temperature source. The refrigerant leaves the evaporator in the form of a vapor and is condensed by an input of work through the compressor. A high pressure is achieved in the compressor, determined by the desired condensing temperature of the refrigerant. The cycle is completed when the refrigerant is returned to the condenser, where heat is rejected to the high-temperature sink.

There are two useful forms of heat-pump cycles: refrigerators and heating systems (hereafter referred to as heat pumps). The objective of a refrigerator is to absorb heat from some refrigerated space, thereby keeping it at a low temperature. A heat pump seeks to deliver heat to a region where a high temperature is desired.

Generally speaking, efficiency is a measure of how much work a system delivers for a given input. From this viewpoint, a refrigerator or a heat pump would have a negative efficiency. Consequently, performance is evaluated by the *coefficient of performance* (COP), which is defined as the ratio of energy sought and the energy that it costs. In the case of a refrigerator, the objective is a high value of heat absorption from the refrigerated space Q_L, while for a heat pump, a high value of heat delivered to the heated space Q_H is sought.

In both cases, the cost of the energy transfer is the work input (W), whose value can be found by applying the first law:

$$W = Q_H - Q_L$$

Therefore, the effectiveness of a refrigerator $(COP)_R$ and a heat pump $(COP)_H$ can be expressed as

$$(COP)_R = \frac{Q_L}{W} = \frac{Q_L}{Q_H - Q_L} = \frac{1}{(Q_H/Q_L) - 1} \tag{8.2}$$

$$(COP)_H = \frac{Q_H}{W} = \frac{Q_H}{Q_H - Q_L} = \frac{1}{1 - (Q_L/Q_H)} \tag{8.3}$$

It follows that for any given system,

$$(COP)_H - (COP)_R = 1.0 \tag{8.4}$$

8.4 ENTROPY

Entropy cannot be measured directly since it is a mathematical abstraction which depends on the functional relationship of certain variables. It has no identity outside of these relationships. Entropy is not really as esoteric a concept as some individuals try to make it. Mathematical formulations for entropy can be used quite easily to achieve quantitative solutions.

One reason why entropy seems so confusing to new students of thermodynamics is that two definitions exist; one is based on macroscopic principles and one on microscopic principles. To a casual observer these definitions might appear to have no similarity. After a certain amount of study, however, it is obvious that the definitions are compatible and complementary.

Macroscopic Entropy

Previous reference was made that there is a preferred sense of direction for all natural processes. When a body falls in the earth's atmosphere, it always falls toward the center of the earth. If an automobile tire is punctured, the enclosed air always flows out of the tire. Human life is conceived, it matures and ages, then it dies.

The processes mentioned above, and many others, always proceed in a direction that is predictable. It occurred to some nineteenth century thinkers that it would be useful if there was a thermodynamic property that would always change in the same direction when a system undergoes a natural change. Clausius, a German physicist, suggested a property called entropy S. He defined the change in entropy for a reversible process in terms of the amount of energy received by a system in the form of heat δQ and the absolute temperature of the system (T):

$$dS \equiv \left[\frac{\delta Q}{T}\right]_{rev} \tag{8.5}$$

or, in the integral form,

$$S_2 - S_1 = \int_1^2 \left[\frac{\delta Q}{T}\right]_{rev} \tag{8.6}$$

In order to solve the integral of Eq. (8.6), some functional relationship between Q and T must be known. In mathematical terms, T is an integrating factor used to convert the inexact differential δQ to the exact differential $[\delta Q/T]$.

The third law of thermodynamics states that the entropy of a pure substance is zero at an absolute temperature of zero. Entropy measured relative to this base is called absolute entropy and is given the symbol $S°$. The concept of absolute entropy is particularly important when chemical reactions are involved. However, when no change in composition occurs, values of entropy can be related to some arbitrarily selected reference point. Engineering steam tables, for example, specify zero entropy at 32.018°F or 0.01°C.

Entropy S is an extensive property, but it is customary to tabulate values of the intensive property specific entropy s. In the saturated region specific entropy can be expressed in terms of tabulated values and quality:

$$\begin{aligned} s &= s_f + xs_{fg} \\ &= (1 - x)s_f + xs_g \\ &= s_g - (1 - x)s_{fg} \end{aligned} \tag{8.7}$$

Microscopic Entropy

Entropy can also be related to the probability that a system will exist in a certain state. This can be done only when a system is made up of many particles so that the laws of probability apply. Such a condition exists when the molecules of a substance are taken as the system. The mathematical formulation for microscopic entropy appears as

$$S = k \ln \Omega \tag{8.8}$$

where k is the Boltzmann constant and Ω is the number of microscopic configurations possible.

Consider, for example, a large tank in which the pressure is less than 1 atm. In one corner of the tank there is a small spherical film which contains a gas at atmospheric pressure. At some instant the film is ruptured, allowing the gas molecules to move freely into the tank. After some period of time we will observe that the gas molecules are uniformly distributed throughout the tank. No matter how long we observe the system, the molecules will never (statistically) return to their original position in the corner. If we call the initial state a "perfectly ordered state," then the entropy is zero since the natural logarithm of one is zero. On the other hand, there are many many ways that the gas molecules can be distributed throughout the tank, and so the entropy (or probability) is high. This means, then, that entropy is a measure of microscopic disorder or randomness in a system.

General Nature of Entropy

In engineering practice the macroscopic concept of entropy is generally used. The microscopic form is used in statistical thermodynamics, and in recent years social scientists have adapted it to communication theory and decision making.

From Eq. (8.5) the units for entropy are given as heat divided by absolute temperature. The units most often used in practice are joule/gm K and Btu/lbm°R.

Certain generalizations can be made concerning the concept of entropy [2,3]:

Entropy is a state property and is associated with all systems.
Total entropy is the sum of all individual entropies.
Entropy can be created but never destroyed.
Entropy is a measure of microscopic disorder and loss of availability.
Entropy of a perfectly ordered system is zero; this occurs at a temperature of absolute zero.
Entropy increases for all irreversible processes.

8.5 STATEMENTS OF THE SECOND LAW

The second law of thermodynamics places limitations on the direction in which a process can proceed and on the efficiency of a process. The most fundamental form of the law states that the total entropy change for a system and its environment is positive for an irreversible process and zero for a reversible process:

$$\Delta S_{total} \geq 0 \tag{8.9}$$

Since all real processes are irreversible, the second law claims that we live in a world of increasing entropy. Many forms of the second law have been proposed, but the two that are most appropriate for thermal technology are the Kelvin-Planck and Clausius statements.

Kelvin-Planck Statement

The Kelvin-Planck statement says that it is impossible to construct a heat engine which operates in a thermal cycle and which produces no effect other than the extraction of heat from a reservoir and the performance of an equivalent amount of work. Figure 8.5a shows the type of device prohibited by the Kelvin-Planck statement. If such a device were possible, an airplane or automobile could be fueled by the earth's atmosphere. Figure 8.5b shows the manner in which any heat engine must operate. A certain amount of heat must be rejected to a sink.

Equation (8.1) gave the thermal efficiency of a heat engine as W/Q_H, where $W = Q_H - Q_L$. The Kelvin-Planck statement implies that $Q_L \neq 0$, so that $Q_H \neq W$. Therefore, an important implication of the Kelvin-Planck statement is that no heat engine can have an efficiency greater than 100%.

Figure 8.5a is referred to as a perpetual motion machine of the second kind (PMM2), and such a device violates the second law. A PMM1 is a device that delivers work without any heat source. This is an obvious violation of the first law. A PMM3 is designated as a device in which there is no friction; it would run forever but would not deliver any work.

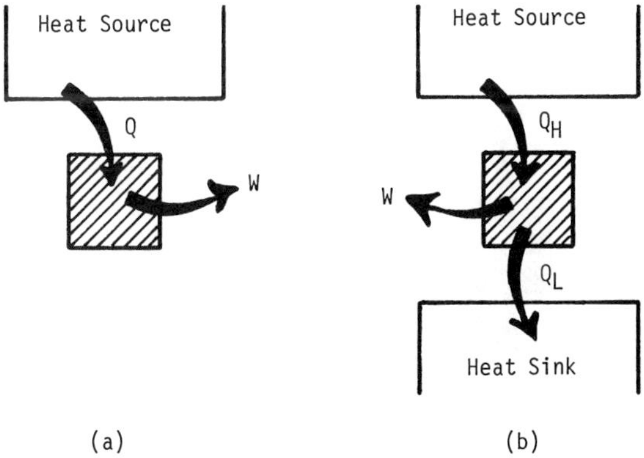

FIGURE 8.5 Illustration of the Kelvin-Planck statement of the second law of thermodynamics: (a) perpetual motion machine of the second kind, prohibited by the second law and (b) all real heat engines have less than 100% efficiency.

Clausius Statement

The Clausius statement says that it is impossible to construct a heat pump that operates in a thermodynamic cycle and which produces no effect other than the transfer of heat from a low temperature source to a high temperature source. Figure 8.6a shows the type of advice prohibited by the Clausius statement. Such a device would operate without any work input, so that $Q_L = Q_H$ from the first law. Recalling Eqs. (8.2) and (8.3), the COP would be infinite for a heat pump or refrigerator since $W = 0$. Figure 8.6b shows the manner in which all heat pumps must operate.

Corollaries to the Second Law

Once the Kelvin-Planck statement of the second law is accepted, many other statements or corollaries can be proved. Some of the more important corollaries are shown below:

Heat will always flow from a high-temperature body to a low-temperature body.

All reversible heat engines (or heat pumps) which operate between the same temperature limits will have the same efficiency (or COP).

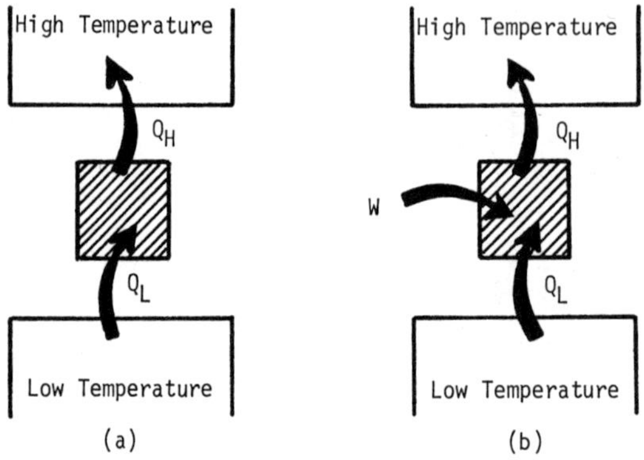

FIGURE 8.6 Illustration of the Clausius statement of the second law of thermodynamics: (a) heat pump without work input--prohibited by the second law, and (b) real heat pump with COP < ∞.

A reversible heat engine (or heat pump) has the maximum possible efficiency (or COP) between any two temperature levels.

An absolute temperature scale can be developed which does not depend on any substance:

$$\left[\frac{Q_H}{Q_L}\right]_{rev} = \frac{T_H}{T_L} \tag{8.10}$$

Certain implications go along with this corollary:

1. No negative absolute temperatures are allowed.
2. Absolute zero temperature can never be reached since $Q_L \neq 0$.

8.6 CARNOT CYCLE

According to the second law of thermodynamics, no cycle can operate at 100% efficiency. It is logical, then, to inquire as to the maximum efficiency that can be obtained. In 1824 Sadi Carnot [4], devised a cycle that is more efficient than any other possible design. The Carnot cycle is made up of four reversible processes, and any working substance amenable to the processes can be used. Figure 8.7 is a schematic illustration of a Carnot heat engine system, where heat is received from a high-temperature reservoir and rejected to a low-

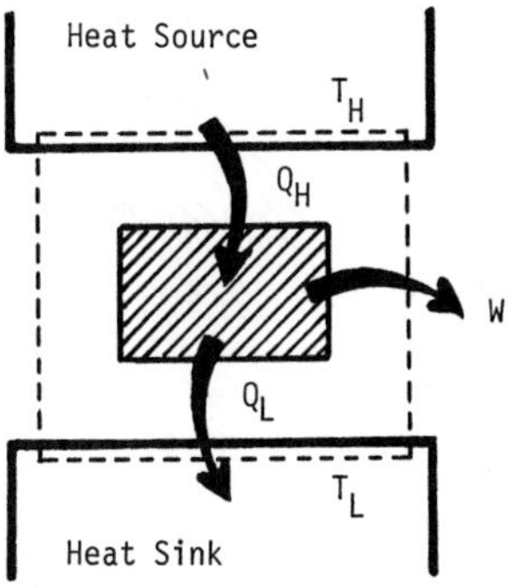

FIGURE 8.7 Schematic illustration of a Carnot engine.

temperature reservoir, resulting in the delivery of positive work. The heat engine can have a variety of forms, and one particular form is similar to that shown in Fig. 8.3.

Useful information about any cycle can be obtained by referring to diagrams of thermodynamic properties for the cycle. Figure 8.8 shows two diagrams for the Carnot heat engine cycle: pressure vs volume and temperature vs entropy. These diagrams are of particular interest since the area under a P-V reversible process curve represents work, and the area under a T-s reversible process curve represents heat transfer. Recalling Eq. (6-11), the net area under the P-V curve represents the net work done by the cycle. On the T-s diagram, area 1-2-b-a-1 represents the heat transferred to the working substance from the high-temperature reservoir, and area 3-b-a-4-3 represents the heat transferred by the working substance to the low-temperature reservoir. Since the net work is equal to the net heat transfer for a Carnot cycle, area 1-2-3-4 indicates net work. Consequently, both diagrams in Fig. 8.8 show the net work done by the engine.

The four reversible processes for a Carnot heat engine can be summarized as follows:

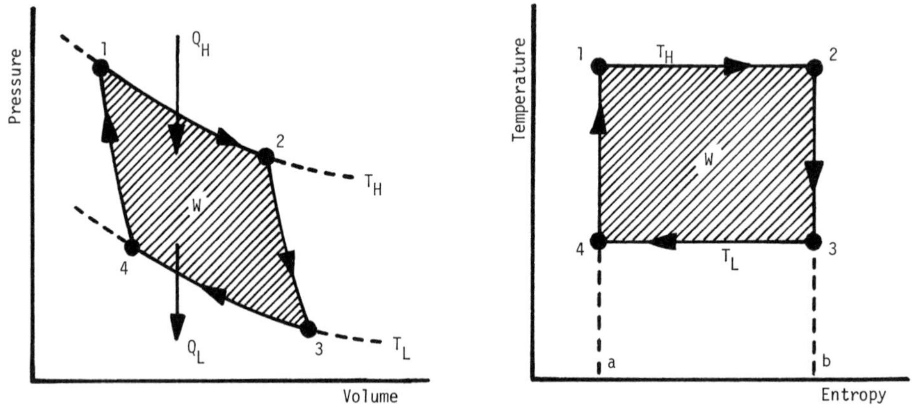

FIGURE 8.8 Carnot heat engine cycle shown on two thermodynamic planes. Area under each curve represents the work done by the cycle.

A reversible isothermal expansion in which the working substance absorbs heat from a high-temperature reservoir.

A reversible adiabatic expansion which lowers the temperature of the working substance from the high value to the low value.

A reversible isothermal compression in which the working substance rejects heat to a low-temperature reservoir.

A reversible adiabatic compression which raises the temperature of the working substance from the low value to the high value.

Several assumptions must be made for the processes in a Carnot cycle to be reversible and for the cycle to attain maximum efficiency. For example, heat is absorbed at a uniform high temperature and rejected at a uniform low temperature since the source and sink are assumed to be thermal reservoirs. The temperature difference between the working substance and the thermal reservoirs--at both the source and sink--must be infinitesimal to prevent any irreversibility. Furthermore, there must be no friction, expansion valves, or explosive processes in any of the devices. Finally, the cycle must operate under steady state conditions so that there is no energy or entropy storage within the system.

In order to obtain an expression for the efficiency of a Carnot cycle, let us make an entropy balance and then apply the second law to Fig. 8.7.

Entropy flows into the system from the high-temperature reservoir and flows out at the low-temperature reservoir:

$$S_{in} = \frac{Q_H}{T_H} \quad \text{and} \quad S_{out} = \frac{Q_L}{T_L}$$

Since there is no entropy flow associated with work, an expression for the total entropy stored in the system is

$$\Delta S_{system} = S_{in} - S_{out} + S_g$$

where S_g is the entropy generated by any irreversibilities in the engine. However, since a Carnot engine operates under steady conditions, there can be no entropy stored in the system and we can express

S_g in terms of the input and output entropies:

$$S_g = \frac{Q_L}{T_L} - \frac{Q_H}{T_H}$$

According to the second law of thermodynamics, entropy must increase or remain the same, which means that

$$\frac{Q_L}{T_L} - \frac{Q_H}{T_H} \geq 0$$

or

$$\frac{Q_L}{Q_H} \geq \frac{T_L}{T_H}$$

The first law can also be applied to Fig. 8.7:

$$W = Q_H - Q_L$$

and, recalling Eq. (8.1):

$$\eta = \frac{W}{Q_H} = \frac{Q_H - Q_L}{Q_H} = 1 - \frac{Q_L}{Q_H} \geq 1 - \frac{T_L}{T_H}$$

The inequality sign applies only to irreversible processes, and since the Carnot cycle has only reversible processes, the equal sign is used. Carnot efficiency is shown to depend only on the absolute temperatures of the reservoirs available:

$$\eta_{Carnot} = 1 - \frac{T_L}{T_H} \tag{8.11}$$

Since all the processes of a Carnot cycle are reversible, the cycle can be operated in the opposite direction as a refrigerator or heat pump. All the statements made concerning a Carnot heat engine apply equally well to a Carnot heat pump; the only change is that performance is measured by coefficient of performance instead of efficiency. Performance evaluation can be obtained for Carnot refrigerators and heat pumps by substituting values of absolute temperature for heat in Eqs. (8.2) and (8.3),

$$(COP)_{Carnot,\ R} = \frac{T_L}{T_H - T_L} \tag{8.12}$$

$$(COP)_{Carnot,\ H} = \frac{T_H}{T_H - T_L} \tag{8.13}$$

8.7 CONCLUSION

In this chapter various statements and corollaries of the second law of thermodynamics are presented. Reference is made to the important concepts of reversibility, irreversibility, availability, and entropy. The second law is shown to place restrictions on the direction in which processes can proceed naturally, and defines the limitations on the performance of heat engines and heat pumps. Efficiency and the coefficient of performance are shown to depend only on the absolute temperature of the thermal reservoirs available.

Entropy is shown to be a mathematical abstraction, which has no identity outside of certain mathematical relations. The distinctions between macroscopic and microscopic entropy are defined, and mathematical formulations for achieving quantitative solutions are given.

The Carnot cycle is introduced as the ideal cycle of maximum efficiency. Necessary assumptions for achieving a Carnot cycle are listed and the four reversible processes that make up the cycle are presented.

SOLVED PROBLEMS

PROBLEM 1: You are asked to comment on an inventor's claim that he has an engine which delivers 25 Btu of work from an input of 100 Btu when it operates between 140°F and 0°F. Is the claim valid?

Solution: The efficiency claimed by the inventor is 25%, which must be compared with the maximum efficiency possible:

$$\eta_t = \frac{T_H - T_L}{T_H} = \frac{600°R - 460°R}{600°R} = 0.233$$

The claim is not valid since the maximum theoretical efficiency possible between the working temperatures is 23.3%.

PROBLEM 2: Saturated steam in a rigid vessel is initially at 10 bar (absolute). Heat is added until the steam is superheated at 38 bar and 310°C. What is the change in entropy?

Solution: Use Keenan and Keyes *Metric Steam Tables*:

At 38 bar and 310°C (table 3)

$$v = 64.03\ \text{cm}^3/\text{gm}$$

$$s = 6.4415\ \text{J/gm K}$$

At 10 bar (table 2)

$$v_f = 1.1273\ \text{cm}^3/\text{gm}$$

$$v_g = 194.44\ \text{cm}^3/\text{gm}$$

$$s_f = 2.1387\ \text{J/gm K}$$

$$s_{fg} = 4.4478\ \text{J/gm K}$$

Since the vessel is rigid, v at 10 bar = 64.03 cm^3/gm

$$v_x = (1 - x)v_f + xv_g$$

$$64.03 = (1 - x)(1.1273) + x(194.44)$$

$$\therefore \quad x = 0.325$$

$$s_x = s_f + xs_{fg}$$

$$s_x = 2.1387 + (0.325)(4.4478) = 3.584\ \text{J/gm K}$$

$$\Delta s = s - s_x = 6.4415 - 3.584 = 2.858\ \text{J/gm K}$$

PROBLEM 3: If heat must be pumped at the rate of 70 000 Btu/hr to keep a house at 72°F when the outside temperature is 15°F, what is the minimum power required to drive the heat pump?

Solution: Minimum power will be required when a Carnot heat pump is used:

$$[\text{COP}]_H = \frac{T_H}{T_H - T_L} = \frac{(72 + 460)}{(72 + 460) - (15 + 460)} = 9.33$$

however

$$[\text{COP}]_H \equiv \frac{\dot{Q}_H}{\dot{W}}$$

$$\therefore \quad \dot{W} = \frac{70\ 000\ \text{Btu/hr}}{9.33} = 7503\ \text{Btu/hr}$$

PROBLEM 4: Heat must be pumped at the rate of 26 kW to maintain a temperature of 43°C in an air stream while the outside temperature is 0°C. Determine (a) the minimum electrical requirement in kilowatts and (b) the electrical requirement if an electric heater is used.

Solution: The minimum power requirement occurs when a Carnot heat pump is used.

(a)

$$(\text{COP})_H = \frac{T_H}{T_H - T_L} = \frac{316\ \text{K}}{316\ \text{K} - 273\ \text{K}} = 7.3$$

$$(\text{COP})_H \equiv \frac{\dot{Q}_H}{\dot{W}} \quad \therefore \quad \dot{W} = \frac{26\ \text{kW}}{7.3} = 3.56\ \text{kW}$$

(b) When an electric heater is used, the heat generated is pure energy dissipation, i.e., $\dot{Q}_H = \dot{W}$:

$$\therefore \quad \dot{W} = 26\ \text{kW}$$

EXERCISES

1. Define the term "exergy."
2. What do all reversible systems have in common?
3. What is the maximum efficiency for an engine working between 127 and 27°C?
4. Based on the usual definition of thermodynamic terms, what is the efficiency of a refrigerator?
5. What is the third law of thermodynamics?
6. Determine the entropy for saturated steam at 25°C and 75% quality.
7. An inventor claims that he has a refrigerator which operates between 0 and 100°C and has a COP of 30. Is the claim valid?
8. Saturated steam in a rigid vessel is initially at 150 psia. Heat is added until the steam is superheated at 550 psia and 550°F. Determine the change in entropy for the process.

9. It is necessary to pump heat at the rate of 15 kW to maintain an enclosure at 20°C when the outside temperature is -7°C. Determine the minimum power that must be supplied to a heat pump that will accomplish the job.
10. What is the change in entropy when 1 g ice at 0°C is converted into 1 g liquid water at 0°C?
11. A kilogram of gas expands adiabatically and reversibly. What is the change in entropy?
12. Determine the change in entropy when 1 g water is heated from 0 to 100°C. Use a thermodynamic calculation and compare the results with values read from steam tables.

REFERENCES

1. G. J. Van Wylen, *Thermodynamics*, Wiley, New York, 1959.
2. G. W. Castellan, *Physical Chemistry*, pp. 152-177, Addison-Wesley, Reading, Mass., 1964.
3. R. E. Balzhiser and M. R. Samuels, pp. 11-13, 38-40, and 136-180, *Engineering Thermodynamics*, Prentice-Hall, Englewood Cliffs, N.J., 1977.
4. S. Carnot, *Reflections on the Motive Power of Heat*, ASME, (translated by R. H. Thurston), 1943.

9

GAS PROPERTIES AND PROCESSES

9.1 INTRODUCTION

Many power systems depend upon the properties of gases or vapors. Recall that the term gas is a general one and that a vapor is a gas near the saturation state.

The behavior of gases or vapors in a system depends on the nature of the process. These processes, in turn, are identified by unique characteristics such as a constant property or no heat flow. Table 9.1 summarizes the most common processes considered in thermodynamics.

9.2 IDEAL GASES

To establish a paragon for all gases, the concept of an ideal gas has been developed. An equation of state can be written for an ideal gas that applies to most gases as the pressure approaches zero, but the equation could not be used near the critical point. For most engineering applications the equation is sufficiently accurate for monatomic and diatomic gases such as air, helium, hydrogen, oxygen, nitrogen, and carbon monoxide. Triatomic gases such as water, carbon dioxide, and sulfur dioxide cannot be treated as ideal gases.

The advantage of the ideal gas equation of state is that it reduces the relationship between thermodynamic properties to a simple

TABLE 9.1 Characteristics of Common Processes

Process	Characteristic or constant property
isobaric isopiestic isopiezic	pressure
isometric isovolumic isochoric	volume
isothermal	temperature
isentropic reversible adiabatic	entropy
isenthalpic	enthalpy
diabatic	with heat flow
adiabatic	without heat flow
polytropic	reversible diabatic

mathematical equation. Real gases are complex and constitute a separate field of study called gas dynamics or kinetic theory [1,2]. A wide variety of equations of state has been developed for real gases, including equations by van der Waals, Dieterici, Berthelot, and Beattie-Bridgeman [3].

Boyle's Law

As long ago as 1680, Robert Boyle measured the volumes of a fixed quantity of gas at different pressures while temperature was kept constant. He found that the PV product was very nearly constant. Boyle's work resulted in a law that bears his name: the volume of a given quantity of gas varies inversely as the pressure when temperature is held constant. Figure 9.1 is a schematic diagram of a piston-cylinder system in which a heat source or sink keeps temperature constant while the pressure and volume of a gas varies. A mathematical statement of the law shows that the pressure-volume relationship is hyperbolic as shown in Fig. 9.1:

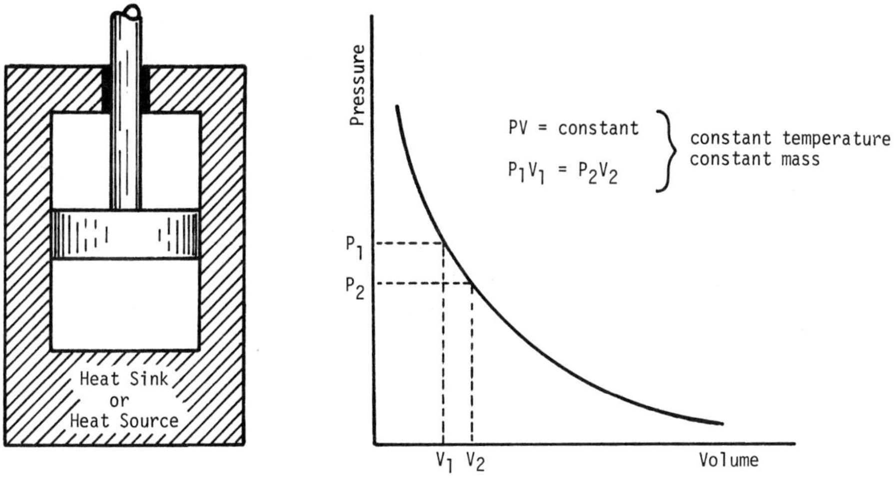

FIGURE 9.1 Boyle's law: the volume of a given quantity of gas varies inversely as the pressure when temperature is held constant.

$$[PV]_{\substack{\text{constant}\\ \text{temp. and mass}}} = \text{constant} \tag{9.1}$$

Boyle's law is also referred to as Mariotte's law.

Charles' Law

Toward the end of the eighteenth century, Jacques Charles found that the pressure of a gas at constant volume changed linearly with temperature. He also observed that the volume of a given quantity of gas was linearly proportional to temperature when pressure was held constant. A little later, Joseph Louis Gay-Lussac came to the same conclusion in independent research, and so the law is variously called Charles' law or Gay-Lussac's law.

Both researchers found that the pressure changes and volume changes were approximately 1/273 of the pressure or volume at 0°C for each degree rise in temperature. Mathematical formulations of this law are summarized below, and schematic illustrations are shown in Figs. 9.2 and 9.3:

$$\left[\frac{P}{T}\right]_{\substack{\text{constant}\\ \text{vol. and mass}}} = \text{constant} \tag{9.2}$$

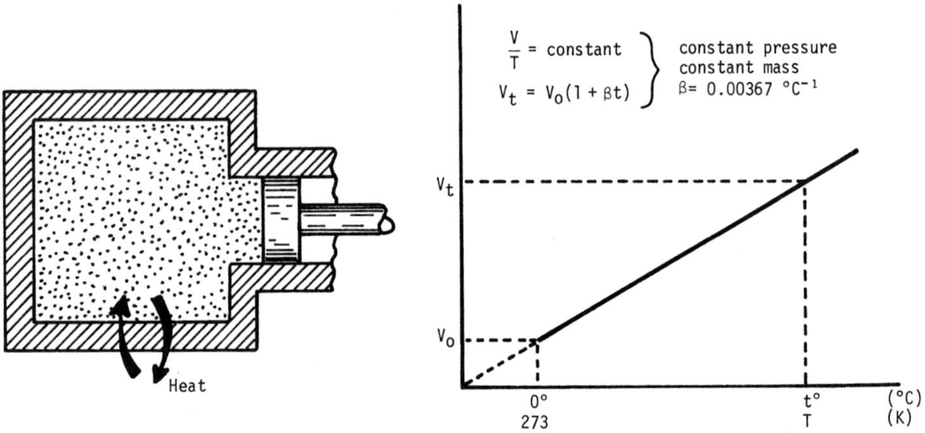

FIGURE 9.2 Charles' (Gay-Lussac) law: the volume of a given quantity of gas is linearly proportional to temperature when the pressure is held constant. (See Fig. 9.3.)

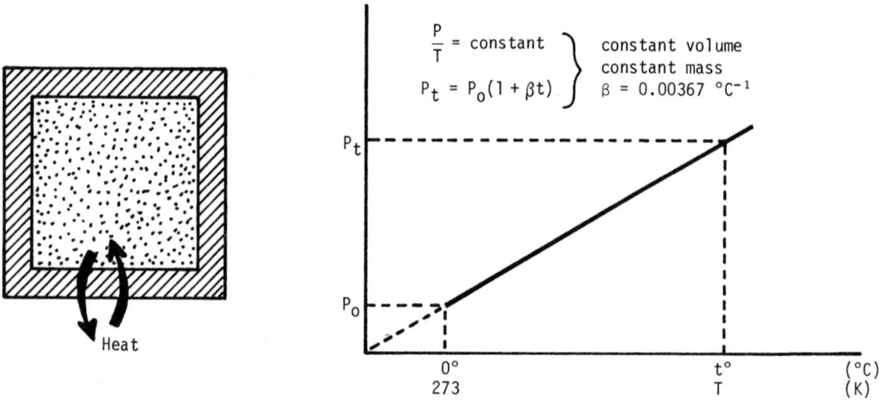

FIGURE 9.3 Charles' (Gay-Lussac) law: The pressure of a given quantity of gas increases (or decreases) by 1/273 of its pressure at 0°C for each 1°C rise (or drop) in temperature when the volume is constant. (See Fig. 9.2).

$$P_t = P_0(1 + \beta t) \tag{9.3}$$

$$\left[\frac{V}{T}\right]_{\substack{\text{constant}\\ \text{press. and mass}}} = \text{constant} \tag{9.4}$$

$$V_t = V_0(1 + \beta t) \tag{9.5}$$

In Eqs. (9.3) and (9.5), P_t and V_t are the pressure and volume at any temperature t, P_0 and V_0 are the pressure and volume at 0°C, and $\beta = 1/273 = 0.00367\ °C^{-1}$ or $0.0022\ °F^{-1}$.

Ideal Equation of State

Consider the vapor or gas phase of a pure substance. A variety of equations of state can be written since a pure substance has only two independent intensive properties. A particular form that is useful for thermodynamic considerations relates pressure (P), specific volume (v), and absolute temperature (T):

$$Pv = RT \tag{9.6}$$

where R is a constant having a unique value for each substance and is called the *gas constant* for the substance. Units for R depend upon those used for pressure, specific volume, and temperature. Table 9.2 shows values of the gas constant for several substances.

TABLE 9.2 Gas Constants for Several Substances

Gas	Molecular Weight	R	
		$\frac{J}{kg\ K}$	$\frac{ft\ lbf}{lbm\ °R}$
Air	28.96	287.1	53.34
Carbon Dioxide	44.01	188.9	35.10
Carbon Monoxide	28.01	296.9	55.16
Helium	4.00	2078.8	386.0
Hydrogen	2.02	4116.3	766.4
Methane	16.04	518.4	96.35
Nitrogen	28.02	296.8	55.15
Oxygen	32.00	259.8	48.28

Conversion of Units

[J/kg K] x (0.186) = [ft lbf/lbm °R]

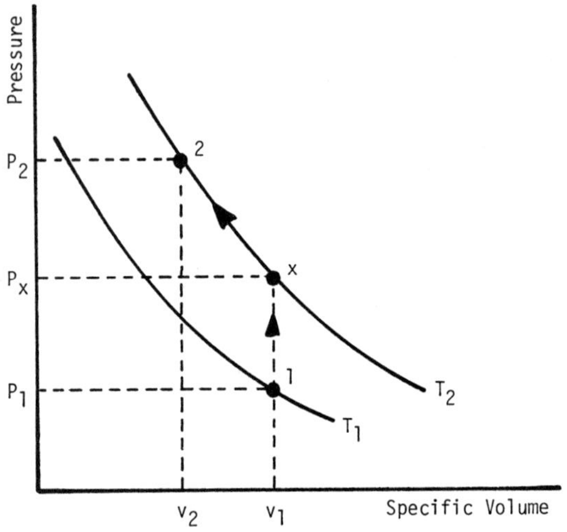

FIGURE 9.4 Ideal gas states on a pressure-specific volume plane.

Equation (9.6) agrees with the laws of Boyle and Charles. Consider, for example, the pressure-specific volume diagram of Fig. 9.4, showing two general states 1 and 2. We can proceed from state 1 to state 2 by an infinite number of paths, but let us go from 1 to x at constant volume, and from x to 2 at constant temperature. From Charles' law:

$$\frac{P_x}{P_1} = \frac{T_x}{T_1} = \frac{T_2}{T_1} \quad \text{or} \quad P_x = P_1 \frac{T_2}{T_1}$$

and from Boyle's law:

$$\frac{P_x}{P_2} = \frac{v_2}{v_x} = \frac{v_2}{v_1} \quad \text{or} \quad P_x = P_2 \frac{v_2}{v_1}$$

Pressure P_x can be eliminated between these relationships and the terms rearranged to give

$$\frac{P_1 v_1}{T_1} = \frac{P_2 v_2}{T_2} = \frac{Pv}{T} = R$$

which is the same form as Eq. (9.6).

It is often convenient to consider gases on a molal basis. A *mole* or *mol* is defined as a mass of substance numerically equal to the molecular weight of the substance. For example, a gram mol of oxygen has a mass of 32 g and a kilogram mol of hydrogen has a mass of 2.02 kg. Obviously, then, the mass (m) of a substance equals the product of the number of mols (n) and the molecular weight M:

$$m = nM \tag{9.7}$$

It follows from this relation that the units of molecular weight can be taken as gm/gm mol, kgm/kgm mol, lbm/lbm mol, etc.

Suppose that Eq. (9.6) is multiplied through by mass m, changing the specific volume to volume:

$$PV = mRT \tag{9.8}$$

and if Eq. (9.7) is applied:

$$PV = nMRT \tag{9.9}$$

Experimental evidence shows that the product MR is constant for all ideal gases; the product is called the *universal gas constant* and is designated $\bar{R}$:

$$\bar{R} \equiv MR \tag{9.10}$$

which results in another form of the ideal gas law:

$$PV = n\bar{R}T \tag{9.11}$$

A constant value of $\bar{R}$ is compatible with Avogadro's law, which states that equal volumes of all gases and vapors at the same temperature and pressure contain the same number of molecules.

Numerical values for $\bar{R}$ depend on the type of units used. Table 9.3 lists values of the universal gas constant that are used in engineering practice. Table 9.4 summarizes several forms of the ideal gas equation of state.

9.3 POLYTROPIC PROCESSES

When a given quantity of gas is expanded or compressed, there are an infinite number of paths it can follow in going from one state to

TABLE 9.3 Universal Gas Constant

$$\bar{R} = 8.315 \times 10^3 \frac{J}{\text{kg mol K}}$$

$$\bar{R} = 8.315 \times 10^7 \frac{\text{ergs}}{\text{g mol K}}$$

$$\bar{R} = 1.986 \frac{\text{cal}}{\text{g mol K}}$$

$$\bar{R} = 0.082 \frac{\text{liter-atm}}{\text{g mol K}}$$

$$\bar{R} = 1545 \frac{\text{ft lbf}}{\text{lbm mol R}}$$

$$\bar{R} = 1.986 \frac{\text{Btu}}{\text{lbm mol R}}$$

$$\bar{R} = 7.805 \times 10^{-4} \frac{\text{hp hr}}{\text{lbm mol R}}$$

$$\bar{R} = 5.819 \times 10^{-4} \frac{\text{kWh}}{\text{lbm mol R}}$$

TABLE 9.4 Several Forms of the Ideal Gas Law

Equation	Number
$Pv = RT$	(9.6)
$PV = mRT$	(9.8)
$PV = nMRt$	(9.9)
$PV = \bar{R}T$	(9.11)

P = pressure

V = volume

v = specific volume

m = mass

n = mols

T = absolute temperature

R = gas constant

$\bar{R}$ = universal gas constant

another. These paths depend on the manner in which heat is absorbed or rejected by the system, and on whether work is done on or by the system. If all the paths are plotted on a P-v plane, a family of hyperbolic curves results with the following generic equation:

$$Pv^n = \text{constant} \tag{9.12}$$

The paths of most interest are those which represent reversible processes. Of these, particular attention is given to four processes in which temperature, pressure, entropy, and volume, respectively, are kept constant. For each of these processes, n has a unique value. Figure 9.5 shows a family of polytropic processes on a P-v plane and Fig. 9.6 shows the four special polytropic processes on P-v and T-s diagrams.

The polytropic exponent n is a constant which must be determined empirically. For example, consider the expansion of combustion gases in the cylinder of a water-cooled reciprocating engine. Pressure and volume at any particular state can be measured by use of an engine indicator. Applying Eq. (9.12) to states 1 and 2:

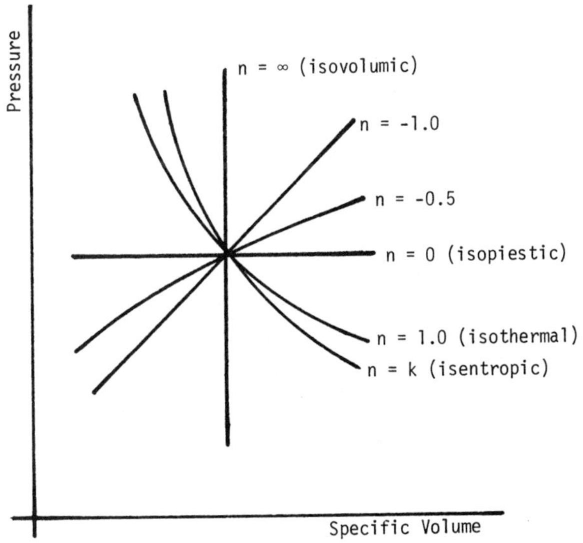

FIGURE 9.5 Family of polytropic processes on a P-v plane.

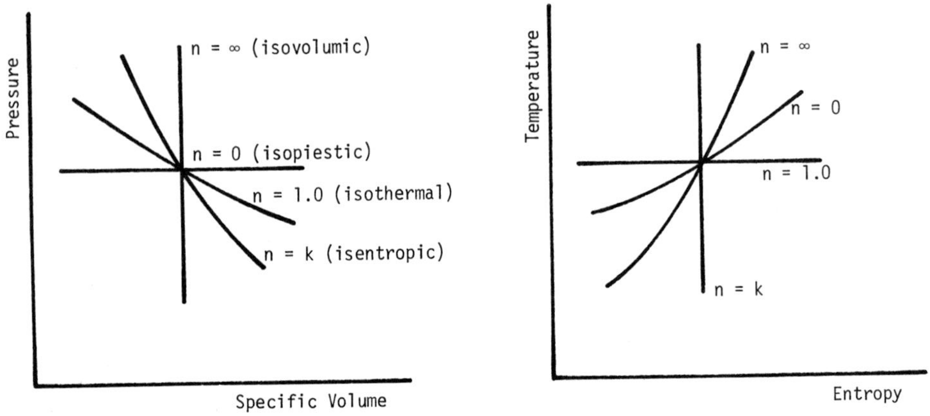

FIGURE 9.6 Four special polytropic processes on P-v and T-s planes.

$$P_1v_1^n = P_2v_2^n$$

Taking the logarithm of both sides, it can be shown that

$$n = \frac{\log P_2 - \log P_1}{\log v_1 - \log v_2} \qquad (9.13)$$

A graphical solution is sometimes more convenient. The logarithmic form of Eq. (9.12) is

$$\log P = -n \log v + \text{constant}$$

which can be converted to the form of a straight line:

$$y = -nx + b$$

in which -n is the slope.

If the general equation for a polytropic process (Pv^n = constant) and the ideal gas law (Pv = RT) are solved simultaneously, several interesting relations among pressure, temperature, and volume can be obtained:

$$\frac{P_2}{P_1} = \left(\frac{v_1}{v_2}\right)^n$$

$$\frac{T_2}{T_2} = \left(\frac{v_1}{v_2}\right)^{n-1} \tag{9.14}$$

$$\frac{T_2}{T_1} = \left(\frac{P_2}{P_1}\right)^{(n-1)/n}$$

Isentropic Process

At ordinary temperatures, specific heats (c_P and c_V) can be assumed to be constant without introducing large errors. If more exact values are needed, the mean specific heat over a given temperature range can be used. The advantage of constant specific heats is that relatively simple relationships can be developed. For an ideal gas with constant specific heat, the following relation holds:

$$c_P - c_V = R \tag{9.15}$$

$$k \equiv \frac{c_P}{c_V} \tag{9.16}$$

The value k is the isentropic exponent that is shown in Fig. 9.6. For an isentropic (reversible adiabatic) process involving an ideal gas with constant specific heat, the general equation for a polytropic process is replaced by the relation

$$Pv^k = \text{constant} \tag{9.17}$$

and a set of relations similar to Eq. (9.14) can be developed:

$$\frac{P_2}{P_1} = \left(\frac{v_1}{v_2}\right)^{k}$$

$$\frac{T_2}{T_1} = \left(\frac{v_1}{v_2}\right)^{k-1} \tag{9.18}$$

$$\frac{T_2}{T_1} = \left(\frac{P_2}{P_1}\right)^{(k-1)/k}$$

Corresponding relationships for each of the other special cases of Fig. 9.6 can be derived. Table 9.5 is a summary of the processes and their relationships. Keep in mind that the relationships apply only to ideal gases with constant specific heat.

Work in Polytropic Processes

Work done by a closed system which undergoes a reversible polytropic process can be found from Eqs. (6.11) and (9.12):

TABLE 9.5 Processes for Ideal Gases with Constant Specific Heats

Process	Pv^n Relation	Final Absolute Temperature (T_2)	Final Specific Volume or Pressure
Polytropic	$n = n$	$T_2 = T_1\left[\frac{v_1}{v_2}\right]^{n-1}$	$v_2 = v_1\left[\frac{P_1}{P_2}\right]^{1/n}$
	$P_1v_1{}^n = P_2v_2{}^n$	$T_2 = T_1\left[\frac{P_2}{P_1}\right]^{n-1/n}$	$v_2 = \frac{RT_2}{P_2}$
		$T_2 = \frac{P_2v_2}{R}$	$P_2 = P_1\left[\frac{v_1}{v_2}\right]^n$
Isentropic	$n = k$	$T_2 = T_1\left[\frac{v_1}{v_2}\right]^{k-1}$	$v_2 = v_1\left[\frac{P_1}{P_2}\right]^{1/k}$
	$P_1v_1{}^k = P_2v_2{}^k$	$T_2 = T_1\left[\frac{P_2}{P_1}\right]^{k-1/k}$	$v_2 = \frac{RT_2}{P_2}$
	$k \equiv \frac{c_p}{c_v}$	$T_2 = \frac{P_2v_2}{R}$	$P_2 = P_1\left[\frac{v_1}{v_2}\right]^k$
Isobaric	$n = 0$	$T_2 = T_1\left[\frac{v_2}{v_1}\right]$	$v_2 = v_1\left[\frac{T_2}{T_1}\right]$
	$P_1 = P_2$	$T_2 = \frac{Pv_2}{R}$	$v_2 = \frac{RT_2}{P}$
Isovolumic	$n = \infty$	$T_2 = T_1\left[\frac{P_2}{P_1}\right]$	$v = \frac{RT_2}{P_2}$
	$v_1 = v_2$	$T_2 = \frac{P_2v}{R}$	
Isothermal	$n = 1$	$T = \frac{Pv}{R}$	$v_2 = v_1\left[\frac{P_1}{P_2}\right]$
	$P_1v_1 = P_2v_2$		$v_2 = \frac{RT}{P_2}$

$$ {}_1W_2 = \int_1^2 P\, dV = \text{constant} \left(\int_1^2 \frac{dV}{V^n} \right) $$

$$ = \frac{\text{constant}}{1 - n} (V_2^{\,1-n} - V_1^{\,1-n}) $$

but

$$ P_1V_1^{\,n} = P_2V_2^{\,n} = \text{constant} $$

$$ \therefore \quad {}_1W_2 = \frac{P_2V_2\, V_2^{\,1-n} - P_1V_1\, V_1^{\,1-n}}{1 - n} $$

$$ {}_1W_2 = \frac{P_2V_2 - P_1V_1}{1 - n} \tag{9.19} $$

and since an ideal gas is being considered, PV = mRT and

$$ {}_1W_2 = \frac{mR(T_2 - T_1)}{1 - n} \tag{9.20} $$

In a similar fashion, an expression can be found for the work done by a steady-flow reversible polytropic process when changes in potential and kinetic energy are negligible. This can be done by combining the relations ${}_1W_2 = -\int v\, dP$ and Pv^n = constant:

$$ {}_1W_2 = -\int_1^2 v\, dP = -\text{constant} \left(\int_1^2 \frac{dP}{P^{1/n}} \right) $$

$$ {}_1W_2 = \frac{n}{n - 1} (P_1v_1 - P_2v_2) \tag{9.21} $$

and since an ideal gas is being considered, Pv = RT and

$$ {}_1W_2 = \frac{nR}{n - 1} (T_1 - T_2) \tag{9.22} $$

Work associated with other types of processes can be found in a similar fashion. The reversible isothermal process is of particular interest because the expression for work is the same for closed and steady flow systems. Also, there is no work associated with a constant volume process. Table 9.6 is a summary of work relationships for several processes.

TABLE 9.6 Work Associated with Ideal Gas Processes with Constant Specific Heat

Process	Work Associated with a Closed System	Work associated with a Steady Flow System
Polytropic	$_1W_2 = \dfrac{P_2V_2 - P_1V_1}{1 - n}$	$_1W_2 = \dfrac{n}{n-1}[P_1v_1 - P_2v_2]$
	$_1W_2 = \dfrac{mR[T_2 - T_1]}{1 - n}$	$_1W_2 = \dfrac{nR}{n-1}[T_1 - T_2]$
	$_1W_2 = \dfrac{P_1v_1}{n-1}\left[1 - \left(\dfrac{v_1}{v_2}\right)^{n-1}\right]$	
Isentropic	$_1W_2 = \dfrac{P_2V_2 - P_1V_1}{1 - k}$	$_1W_2 = \dfrac{k}{k-1}[P_1v_1 - P_2v_2]$
	$_1W_2 = \dfrac{mR[T_2 - T_1]}{1 - k}$	$_1W_2 = \dfrac{kR}{k-1}[T_1 - T_2]$
	$_1W_2 = \dfrac{P_1v_1}{k-1}\left[1 - \left(\dfrac{v_1}{v_2}\right)^{k-1}\right]$	
Isothermal	$_1W_2 = P_1V_1 \log_e \dfrac{V_2}{V_1} = mRT \log_e \dfrac{V_2}{V_1}$	
	$_1W_2 = P_1V_1 \log_e \dfrac{P_1}{P_2} = mRT \log_e \dfrac{P_1}{P_2}$	
Isobaric	$_1W_2 = P(v_2 - v_1)$	0
	$_1W_2 = R(T_2 - T_1)$	

9.4 GAS AND VAPOR MIXTURES

Many examples can be cited in engineering where mixtures of gases or mixtures of gases and vapors must be considered: atmospheric air, products of combustion, fuel mixtures, etc. When the mixtures are homogeneous and purely physical, and when no chemical reactions occur, properties of the mixture can be predicted from properties of the constituents:

The pressure of a mixture of ideal gases is equal to the sum of the partial pressures of the components:

$$P_m = P_a + P_b + P_c + \cdots \tag{9.23}$$

This law is called Dalton's law for partial pressures. It holds exactly for an ideal gas and has a high degree of accuracy even when the mixture deviates from ideal gas conditions.

The internal energy of a mixture of gases equals the sum of internal energies of the components at the pressure and temperature of the mixture:

$$U_m = U_a + U_b + U_c + \cdots \tag{9.24}$$

The entropy of a mixture of ideal gases equals the sum of entropies of the components. The entropy of each of the component gases must be taken at its partial pressure:

$$S_m = S_a + S_b + S_c + \cdots \tag{9.25}$$

The volume of a mixture of ideal gases is equal to the sum of the partial volumes of the components (Amagat's law):

$$V_m = V_a + V_b + V_c + \cdots$$

and if the mixture is homogeneous, the absolute temperature of the components will be equal:

$$T_m = T_a = T_b = T_c = \cdots$$

and from the principle of conservation of mass:

$$m_m = m_a + m_b + m_c + \cdots$$

The gas constant for the mixture can be found with the aid of Eq. (9.10):

$$R_m = \frac{\bar{R}}{M_m}$$

where $\bar{R}$ is the universal gas constant and M_m is the apparent molecular weight of the mixture. Since

$$M_m = M_a + M_b + M_c + \cdots$$

therefore

$$R_m = \frac{M_a R_a + M_b R_b + M_c R_c + \cdots}{M_a + M_b + M_c + \cdots} \qquad (9.26)$$

Considering the relationships shown above, an equation of state can be written for an ideal gas mixture:

$$P_m V_m = m_m R_m T_m \qquad (9.27)$$

9.5 CONCLUSION

In this chapter the definition of an ideal gas and its simplified equation of state are presented. The dependence of the law on basic laws stated by Boyle, Charles, Gay-Lussac, and Marriotte is demonstrated. The concept of the universal gas constant is introduced.

When a given quantity of gas is expanded or compressed, it is shown that an infinite number of paths can be followed. Four particular paths comprising the polytropic processes are described, and the polytropic exponent n is defined. Mathematical formulations for analyzing polytropic processes are shown in tabular form.

Rules are presented for determining the properties of mixtures of gases and vapors from properties of the constituents.

SOLVED PROBLEMS

PROBLEM 1: A 15 ft^3 volume of an ideal gas has a mass of 20 lbm at 80°F and 320 psia. What is the molecular weight of the gas?

Solution:

$$PV = mRT$$

$$R = \frac{[320\ \text{lbf/in}^2][144\ \text{in}^2/\text{ft}^2][15\ \text{ft}^3]}{[20\ \text{lbm}][540°\text{R}]} = 64\ \text{ft lbf/lbm °R}$$

but

$$R = \frac{\bar{R}}{M}$$

$$\therefore \quad M = \frac{1545 \text{ ft lbf/lbm mol}^\circ\text{R}}{64 \text{ ft lbf/lbm}^\circ\text{R}} = 24.1 \text{ lbm/lbm mol}$$

PROBLEM 2: A 50 liter tank of oxygen is initially at 20 atm and 30°C. A sufficient amount of oxygen is removed to change the state to 6 atm and 10°C. Determine (a) the initial mass of oxygen and (b) the final mass of oxygen.

Solution:

(a)

$$pV = n\bar{R}T$$

$$\therefore \quad n = \frac{[20 \text{ atm}][50 \ \ell]}{[0.082 \ \ell \text{ atm/gm mol K}][303 \text{ K}]} = 40.24 \text{ gm mol}$$

$$m = nM \quad \therefore \quad m_1 = (40.24 \text{ gm mol})(32 \text{ gm/gm mol}) = 1288 \text{ gm}$$

(b)

$$n_2 = \frac{(6)(50)}{(0.082)(283)} = 12.9 \text{ gm mol}$$

$$m_2 = (12.9)(32) = 414 \text{ gm}$$

PROBLEM 3: If 0.75 ft^3 of a gas at 1 atm and 35°F is compressed polytropically to a pressure of 3 atm, determine the final volume and temperature. The polytropic exponent is 1.3.

Solution:

$$P_1V_1^n = P_2V_2^n$$

$$V_2^n = \frac{(1.0 \text{ atm})(0.75 \text{ ft}^3)^{1.3}}{(3 \text{ atm})} = 0.229$$

$$\therefore \quad V_2 = (0.229)^{1/1.3} = 0.32 \text{ ft}^3$$

Assuming the gas to be ideal:

$$\frac{P_1V_1}{T_1} = \frac{P_2V_2}{T_2}$$

$$\therefore \quad T_2 = \frac{(3 \text{ atm})(0.32 \text{ ft}^3)(495^\circ\text{R})}{(1.0 \text{ atm})(0.75 \text{ ft}^3)} = 634^\circ\text{R}$$

PROBLEM 4: A rigid container holds a 0.45 kg mass of air with a volume of 0.03 m^3 and an absolute pressure of 6.9×10^5 Pa. The air is expanded reversibly and isothermally to twice its initial volume. It is then compressed reversibly and adiabatically to 0.03 m^3. The air is then returned to its initial state by a reversible constant volume process. Determine the pressure and temperature at each state and show the cycle on a PV and Ts diagram (k = 1.40 for air).

Solution:

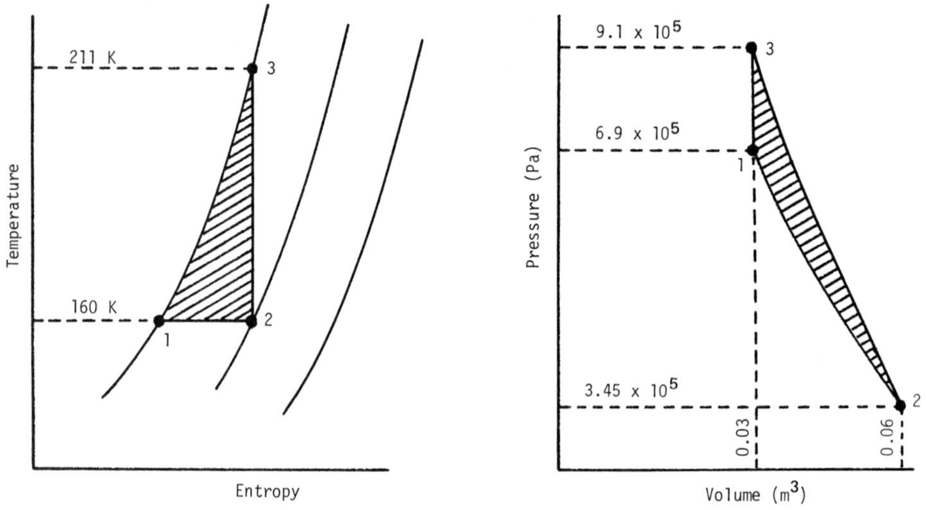

For state 1:

$$PV = mRT$$

$$\therefore \quad T_1 = \frac{(6.9 \times 10^5\ N/m^2)(0.03\ m^3)}{(0.45\ kg)(287.1\ J/kg\ K)} = 160\ K$$

For state 2 (isothermal process, therefore, PV = constant):

$$P_2 = P_1 \frac{V_1}{V_2} = 6.9 \times 10^5 \frac{0.03}{0.06} = 3.45 \times 10^5\ Pa$$

also

$$T_2 = T_1 = 160 \text{ K}$$

For state 3:

$$\left.\begin{aligned} \frac{T_3}{T_2} &= \left(\frac{V_2}{V_3}\right)^{k-1} \\ \frac{P_3}{P_2} &= \left(\frac{V_2}{V_3}\right)^{k} \end{aligned}\right\} \quad \text{isentropic process}$$

$$T_3 = (160)\left(\frac{0.06}{0.03}\right)^{1.4-1.0} = 211 \text{ K}$$

$$P_3 = (3.45 \times 10^5 \left(\frac{0.06}{0.03}\right)^{1.4} = 9.1 \times 10^5 \text{ Pa}$$

PROBLEM 5: Nitrogen is compressed in a reversible polytropic process from 1 atm and 60°F to 4 atm. Calculate the work associated with the process if the polytropic exponent is 1.3.

Solution:

$$T_2 = T_1 \left(\frac{P_2}{P_1}\right)^{(n-1)/n} = (520°\text{R})\left(\frac{4 \text{ atm}}{1 \text{ atm}}\right)^{1.3-1.0/1.3} = 716°\text{R}$$

For a closed system:

$$\frac{W}{m} = \frac{°R(T_2 - T_1)}{1 - n} = w$$

$$W = \frac{55.15 \text{ ft lbf/lbm°R } (716\text{-}520)°\text{R}}{1 - 1.3} = -36031 \text{ ft lbf/lbm}$$

$$W = \frac{-36\ 031}{778} = -46.3 \text{ Btu/lbm}$$

(-) signifies that work must be supplied to the system.

PROBLEM 6: Determine the total pressure of a mixture of 10 lbm oxygen and 15 lbm nitrogen at 120°F in a rigid tank of 150 ft^3. Show that the total volume is the sum of the partial volumes.

Solution: Assume O_2 and N_2 to be ideal gases:

$$PV = mRT$$

$$P_{O_2} = \frac{(10\ \text{lbm})(48.28\ \text{ft lbf/lbm}^\circ\text{R})(580^\circ\text{R})}{150\ \text{ft}^3} = 1867\ \text{lb/ft}^2$$

$$P_{N_2} = \frac{(15)(55.15)(580)}{150} = 3199\ \text{lb/ft}^2$$

From Dalton's law:

$$P_{\text{mixture}} = 1867 + 3199 = 5066\ \text{lb/ft}^2$$

Partial volumes can also be found by means of the ideal gas equation:

$$V_{O_2} = \frac{(10\ \text{lbm})(48.28\ \text{ft lbf/lbm}^\circ\text{R})(580^\circ\text{R})}{5066\ \text{lb/ft}^2} = 55.3\ \text{ft}^3$$

$$V_{N_2} = \frac{(15)(55.15)(580)}{5066} = 94.7\ \text{ft}^3$$

$$\therefore \quad V_{\text{mixture}} = 55.3 + 94.7 = 150\ \text{ft}^3 \qquad \text{Q.E.D.}$$

EXERCISES

1. A container holds 500 ml of gas at 20°C and 1 atm pressure. If the pressure is doubled while the temperature is held constant, how does the volume change?
2. A gas at 50°C has a volume of 250 ml. What is the volume at 0°C if the pressure is held constant?
3. A room 18 × 20 × 8 ft contains air at 70°F and standard atmospheric pressure. What is the mass of the air?
4. If 1g gas at 180°C and 0.7 atm occupies 0.5 liter, what is the molecular weight of the gas?
5. An automobile tire has a gauge pressure of 1.9 atm and a temperature of 70°F when it is stationary. After moving a certain distance the tire pressure increases to 2.1 atm gauge. What is the corresponding temperature?
6. A 15 ft^3 volume of nitrogen at 13 atm absolute and 200°F is cooled until the temperature is 70°F. Determine (a) the amount of heat removed and (b) the final pressure.

7. Suppose 3 mol oxygen at 4 atm absolute pressure and 450°F are contained in a compartment which is valve-connected to a compartment with 4 mol nitrogen at 2 atm absolute pressure and 120°F. What is the equilibrium temperature if the valve is suddenly opened and adiabatic mixing occurs?
8. Suppose 1 ft^3 of air is compressed isentropically from 15 psia and 45°F to 60 psia. If the polytropic exponent for air is 1.4, determine the final volume.
9. A volume of air is halved by a reversible adiabatic compression. If the initial state is 270°R and 50 psia, what is the final state?
10. Suppose 1 ft^3 of air with a mass of 1 lb is doubled in volume by a reversible isothermal expansion. If the initial pressure is 100 psia, determine the final pressure and temperature.
11. Methane is compressed in a reversible polytropic process from 1 atm and 15°C to 3 atm. If the polytropic exponent is 1.3, determine the work associated with the process.
12. A tank of 4 m^3 volume holds a mixture of 5 kg of oxygen and 8 kg of nitrogen at 50°C. What is the pressure of the mixture?

REFERENCES

1. J. H. Hildebrand, *An Introduction to Molecular Kinetic Theory*, Reinhold, New York, 1963.
2. J. Jeans, *An Introduction to the Kinetic Theory of Gases*, Cambridge University Press, Cambridge, England, 1962.
3. G. W. Castellan, *Physical Chemistry*, Addison-Wesley, Reading, Mass., 1964.

10

COMBUSTION PROCESSES

10.1 INTRODUCTION

When certain substances are combined, chemical reactions occur and the result is a liberation or absorption of chemical energy. Each chemical reaction is accompanied by a definite change in the heat content of the system. In the large majority of reactions, heat is generated, and these reactions are called *exothermic*. For a few reactions such as the oxidation of nitrogen to form nitric oxide, heat is absorbed; these reactions are called *endothermic*.

In thermal technology an important chemical reaction occurs in a combustion process. Combustion is the rapid combination of a fuel with oxygen, which is accompanied by the liberation of heat and light. Three requirements are necessary for a combustion process to occur:

Thorough mixing of fuel and oxygen in a proportion that will cause complete combustion.

Exposure of fuel particles to oxygen for a period of time that is sufficient to cause complete combustion.

Maintenance of a temperature above the ignition temperature of the fuel.

The ignition temperature is defined as the lowest temperature which will cause a flame to start and spread throughout a combustible

TABLE 10.1 Ignition Temperatures

Substance	Temp., °C
Carbon	718
Carbon monoxide	1218
Heptane	1058
Hydrogen	438
Iso-octane	948
Sulfur	468

mixture. Table 10.1 shows values of ignition temperatures for several substances.

10.2 FUELS

Fuels are generally classified under three headings: gaseous, liquid, or solid. Table 10.2 shows examples of each classification.

The combustible elements that characterize commercial fuels are carbon, hydrogen, and sulfur. When a unit mass of a combustible combines with a sufficient amount of oxygen, the amount of heat generated is called the *heat of combustion* or the *heat value* of the fuel. These values are determined experimentally through the use of calorimeters and can be found in a variety of publications [1,2]. Table 10.3 lists values of heats of combustion for a number of fuels of technical interest.

10.3 COMBUSTION EQUATIONS

During a combustion process the mass of each element is conserved; consequently, chemical equations can be written which equate the quantity of reactants to the quantity of products. Products of a combustion process are carbon dioxiae, water, and sulfur dioxide. If the oxygen source is air, then nitrogen and excess oxygen are also among the products of combustion. The presence of carbon monoxide as a product indicates incomplete combustion.

TABLE 10.2 Classification of Fuels

Natural Gas	pumped from underground reservoirs	Gaseous Fuels
Manufactured Gas	made from coal, coke, oil, natural gas	Gaseous Fuels
Mixed Gas	combination of natural and manufactured gas	Gaseous Fuels
Liquefied Petroleum Gas	propane butane	Gaseous Fuels
Gasoline	complex mixture of hydrocarbons	Liquid Fuels
Kerosene	less volatile than gasoline	Liquid Fuels
Diesel Oil Fuel Oils	petroleum distillates	Liquid Fuels
Alcohol	Ethyl C_2H_5OH (grain) Methyl CH_3OH (wood)	Liquid Fuels
Coal		Solid Fuels
Coke		Solid Fuels
Wood		Solid Fuels
Rocket Fuels	nitrocellulose nitroglycerin	Solid Fuels

The word *stoichiometry* [3] comes from Greek and signifies the mass relations of chemical elements. A *stoichiometric equation* is one in which the masses of all elements are balanced. Consider, for example, the oxidation of carbon:

$$\underset{\text{reactants}}{C + O_2} \rightarrow \underset{\text{product}}{CO_2}$$

This equation states that 1 mol carbon reacts with 1 mol oxygen to form 1 mol carbon dioxide. The molecular weights of the reactants are, 12 and 32 respectively, while the molecular weight of the product is 44. From the definition of a mol, the equation also states

TABLE 10.3 Heats of Combustion

Substance	Equation	State	J/g
Hydrogen	H_2	gas	141 863
Methane	CH_4	gas	55 145
Acetylene	C_2H_2	gas	50 208
Ethane	C_2H_6	gas	51 380
Propylene	C_3H_6	gas	48 831
Propane	C_3H_8	gas	50 044
Benzene	C_6H_6	liquid	41 961
n Hexane	C_6H_{14}	liquid	48 154
Toluene	C_7H_8	liquid	42 484
n Octane	C_8H_{18}	liquid	47 811
Napthalene	$C_{10}H_8$	solid	40 288

Unit Conversion

[J/g] x (0.430) = [Btu/lbm]

[J/g] x (0.239) = [cal/g]

[Btu/lbm] x (0.556) = [cal/g]

that 12 kg carbon combines with 32 kg oxygen to form 44 kg carbon dioxide; other mass units can also be used. Since the mass on each side of the equation is balanced, the equation is stoichiometric.

We can also consider the oxidation of hydrogen:

$$H_2 + \frac{1}{2} O_2 \rightarrow H_2O$$

Generally speaking, 1 mol fuel is used in combustion equations, but to eliminate the use of fractional coefficients, a variation of the stoichiometric equation can be used:

$$2\ H_2 + O_2 \rightarrow 2\ H_2O$$

which indicates that 2 mol hydrogen and 1 mol oxygen produce 2 mol water.

When a hydrocarbon fuel such as methane (CH_4) is burned, both the carbon and hydrogen are oxidized:

$$\underset{\text{reactants}}{CH_4 + 2\ O_2} \rightarrow \underset{\text{products}}{CO_2 + 2\ H_2O}$$

10.4 AIR REQUIRED FOR COMBUSTION

For most combustion processes, oxygen is supplied in the form of atmospheric air rather than as pure oxygen. A volumetric analysis of air shows that it consists primarily of oxygen and nitrogen, with small amounts of argon and carbon dioxide and traces of other gases. For engineering calculations air is considered to be made up of 21% oxygen and 79% nitrogen on a volume basis (23.2% and 76.8% on a mass basis).

Air that contains the minimum amount of oxygen required to complete a combustion process is called *theoretical air*. For example, the following relation indicates the burning of methane in the presence of theoretical air:

$$CH_4 + 2\ O_2 + 2(3.76)\ N_2 \rightarrow CO_2 + 2\ H_2O + 7.52\ N_2$$

Note that the nitrogen is not involved in the combustion process and consequently does not change. In practice it has been found that theoretical air does not result in complete combustion. Consequently, it is customary to specify an excess amount of air for all industrial combustion. Let us consider, for example, the burning of methane in the presence of 150% theoretical air:

$$CH_4 + 2(1.5)\ O_2 + 2(1.5)(3.76)\ N_2 \rightarrow CO_2 + 2\ H_2O + O_2 + 11.28\ N_2$$

An important parameter for combustion processes is the air-fuel ratio (A/F). The ratio can be expressed on a molal or a mass basis. Consider, for example, the air-fuel ratio required to burn methane with theoretical air. From the equation for this process, 1 mol fuel requires 9.52 mol air:

$$\therefore \quad [A/F]_{theor} = 9.52\ \frac{\text{mol air}}{\text{mol fuel}}$$

The result can be obtained on a mass basis by recalling that the molecular weights of air and methane are 28.96 and 16 (Table 9.2):

$$\therefore \quad [A/F]_{theor} = \frac{(9.52)(28.96)}{16} = 17.23 \frac{\text{kg air}}{\text{kg fuel}}$$

10.5 CONCLUSION

Chemical reactions occur in many thermal systems and the most common example is a combustion process. This chapter covers the requirements necessary for a combustion process to occur. Classification of fuels are outlined and combustion equations for stoichiometric processes are developed. Methods are outlined for determining the air-fuel ratio required for complete combustion.

SOLVED PROBLEMS

PROBLEM 1: If 3 mol carbon are completely burned in a process, determine (a) the number of moles of O_2 required and (b) the mass of O_2 required.

Solution:

(a) $3\ C + 3\ O_2 \rightarrow 3\ CO_2$, therefore, 3 mol O_2 are required.

(b)

$$m = Mn$$

where

m = mass of O_2

M = 32 lbm/lbm mol

n = 3 lbm mol

$m = (32)(3) = 96$ lbm

PROBLEM 2: Develop the stoichiometric equation that expresses the complete combustion of ethane (C_2H_6).

Solution: The equation can be developed from the basic relations showing the oxidation of carbon and hydrogen. One mole of ethane contains 2 mol C and 3 mol H_2:

$$2\ C + 2\ O_2 \rightarrow 2\ CO_2$$

$$3H_2 + \frac{3}{2} O_2 \rightarrow 3\ H_2O$$

These equations can be added:

$$C_2H_6 + \frac{7}{2} O_2 \rightarrow 2\ CO_2 + 3\ H_2O$$

PROBLEM 3: Determine the air-fuel ratio on a gram-molal and a mass basis if octane is completely burned in the presence of theoretical air.

Solution: Octane (C_8H_{18}) is made up of 8 mol carbon and 9 mol hydrogen. The stoichiometric equation can be developed from the basic relations showing the oxidation of carbon and hydrogen.

$$8\ C + 8\ O_2 \rightarrow 8\ CO_2$$

$$9\ H_2 + \frac{9}{2} O_2 \rightarrow 9\ H_2O$$

Adding these equations:

$$C_8H_{18} + \frac{25}{2} O_2 \rightarrow 8\ CO_2 + 9\ H_2O$$

There are 25/2 mol oxygen required for complete combustion, or

$$\frac{25}{2} \text{ mol } O_2 + \frac{25}{2} (3.76) \text{ mol } N_2 \rightarrow 59.5 \text{ mol air}$$

$$\therefore \quad (A/F)_{theor} = 59.5 \text{ mol air/mol fuel}$$

Molecular weight of C_8H_{18} = 114 gm/gm mol

Molecular weight of air = 28.96 gm/gm mol

$$\therefore \quad (A/F)_{theor} = 59.5 \frac{(28.96)}{114} = 15.1 \text{ gm air/gm fuel}$$

PROBLEM 4: Develop the stoichiometric equation that expresses the process of octane being burned with 150% theoretical air. What is the air-fuel ratio on a kilogram molal basis?

Solution: From solved problem 3, 12.5 mol O_2 is required for the combustion of octane with theoretical air.

$$C_8H_{18} + (12.5)(1.5)O_2 + (12.5)(1.5)(3.76)N_2 \rightarrow 8\ CO_2 + 9\ H_2O + 1.5\ O_2 + 70.5\ N_2$$

$$\therefore \quad 18.75 \text{ mol } O_2 + 70.5 \text{ mol } N_2 \rightarrow 89.25 \text{ mol air}$$

$$(A/F)_{150\%} = 89.25 \text{ kg mol air/kg mol fuel}$$

PROBLEM 5: Benzene is used in a heating process that is 65% efficient. How much benzene is required to heat a 185 liter tank of water from 10° to 30°C?

Solution: The amount of heat required to raise the temperature of water can be found from the relation

$$Q = mc\ \Delta t$$

where

Q = amount of heat

$m = (185\ell)(10^3 \text{ cm}^3/\ell)(1 \text{ gm/cm}^3) = 1.85 \times 10^5 \text{ gm}$

$c = 4.19 \text{ J/gm °C}$

$\Delta t = 30 - 10 = 20°C$

$$Q = (1.85 \times 10^5)(4.19)(20) = 1.55 \times 10^7 \text{ J}$$

From Table 10.3, the heat of combustion for benzene is 41 961 J/gm.

$$(1.55 \times 10^7 \text{ J}) = (m_{C_6H_6})(41\ 961 \text{ J/gm})(0.65)$$

$$m_{C_6H_6} = 568 \text{ gm}$$

EXERCISES

1. Define (a) combustion, (b) ignition temperature, and (c) heat of combustion.
2. When one "blows out" a match or candle, what actually causes the combustion to cease?
3. Determine the number of moles and the mass of oxygen required to completely burn 5 g mol methane.
4. Write the stoichiometric equation for the complete combustion of iron to form ferric oxide (Fe_2O_3).

5. Develop the relationship that expresses the complete combustion of pentane (C_5H_{12}).
6. For engineering considerations air is made up of 21% oxygen and 79% nitrogen by volume; therefore, for each mole of oxygen there are 79/21 = 3.76 mol nitrogen.

$$\therefore \quad 1 \text{ mol } O_2 + 3.76 \text{ mol } N_2 \rightarrow 4.76 \text{ mol air}$$

   Write a corresponding relationship on a mass basis.
7. Determine the air-fuel ratio on a pound molal and pound mass basis when benzene is completely burned in the presence of theoretical air.
8. Develop the stoichiometric equations for the complete combustion of toluene in the presence of 200% theoretical air. What is the air-fuel ratio?
9. If a heating process is 70% efficient, how much napthalene is required to raise the temperature of 1 liter of water 10°C?
10. An 8-lb mass of fuel oil is required in a 65% efficient process to heat 60 gal water from 50 to 140°F. What is the heat of combustion for the fuel oil?

REFERENCES

1. *Handbook of Physics and Chemistry*, 49th ed., pp. D-184 to D-189, Chemical Rubber To., Cleveland, Ohio, 1969.
2. *Marks Standard Handbook for Mechanical Engineers*, 7th ed., pp. 4-69 to 4-79, McGraw-Hill, New York, 1969.
3. M. J. Sienko and R. A. Plane, 3rd ed., *Chemistry*, McGraw-Hill, New York, 1966.

11

CONDUCTION HEAT TRANSFER

11.1 INTRODUCTION

Thermal conduction is a process by which heat flows through a solid, liquid, or gaseous medium, or between two media which are in intimate contact. In accordance with the second law of thermodynamics, heat flows from a higher to a lower temperature level.

The mechanism of conduction can be studied from two viewpoints: the microscopic and macroscopic. The conclusions reached are not exactly the same because the former depends on statistical thermodynamics while the latter is based on classical thermodynamics. Engineers are primarily interested in the macroscopic approach because it is the phenomenon whose effects are directly transmitted to equipment. In addition, the mathematical treatment for macroscopic conduction has been well developed and is relatively simple to use.

Even though macroscopic techniques are used to analyze conduction processes, a knowledge of the microscopic characteristics of materials is important for a qualitative understanding of conduction. Such a knowledge requires that the solid, liquid, and gas phases of materials be considered separately [1-3].

Conduction in Solids

Materials science distinguishes between dielectric solids and metallic or conductive solids. In the case of dielectric solids, conduction is almost entirely due to lattice vibrations, while for metallic solids there is an added energy transport by free electrons. Thermal energy is transported by free electrons in much the same manner as electric charge. It is not surprising, therefore, that good electrical conductors, such as silver and copper, are also good heat conductors.

Conduction in Liquids

After a change in phase from solid to liquid, there is a lessening of intermolecular bonds and a disruption of the ordered state of the solid results. More freedom for thermal motion of the molecules is available, and so the contact areas of the molecules, on the average, are farther apart and the thermal conductivity is lowered. Liquids are more transparent than solids, and so it is sometimes difficult to distinguish between the thermal energy transported by conduction and by radiation. This characteristic accounts for the difficulty in making measurements of the thermal conductivity of liquids.

Conduction in Gases

During an evaporative process in which a liquid changes to gas, intermolecular bonds are loosened still further. Gas molecules are free to move in any direction, and the only restraints are random collisions. In order to analyze the molecular motion, the kinetic theory of gases must be used. The random motion, characteristic of gas molecules, results in a much lower probability that contact areas will exist, and the thermal conductivity is quite small.

11.2 FUNDAMENTAL LAW

In 1822, Fourier postulated the conduction law for heat transfer, based on experimental data:

Quantity of heat per unit time $\propto$ (normal area) (temperature gradient)

The proportionality constant k, which is needed to replace the proportionality sign with an equal sign, is called the thermal conductivity coefficient:

$$\frac{\partial Q}{\partial \tau} = -kA \frac{\partial t}{\partial x} \tag{11.1}$$

The negative sign is assigned arbitrarily so that the heat flow can be considered positive when the temperature decreases along the path. In this equation Q is the amount of heat, τ is time, A is the area of the conduction medium normal to the heat flow, t is temperature, and x is the length measurement along the heat path.

The partial differential form of Eq. (11.1) indicates that Q and t are functions of space and time coordinates. This is characteristic of *transient conduction*, which can be contrasted with *steady state conduction*. In the latter case, heat and temperature are functions of space coordinates only.

11.3 THERMAL CONDUCTIVITY

Thermal conductivity is a mathematical constant that relates heat transfer through some area to the temperature gradient which induces the flow. However, it is also a physical property of a material and depends on the chemical composition, the physical structure, and the state of the material in question. For example, the thermal conductivity of a solid is generally higher than that of a liquid or gas, and porous materials usually exhibit less conductivity than materials of high density. Grained materials are often anisotropic, i.e., the thermal conductivity depends on the direction in which it is measured. The most striking examples are certain types of wood, such as oak and pine, where the conductivity parallel to the grain is two to four times the conductivity perpendicular to the grain.

The purity of a material can have a significant influence on its thermal conductivity; this characteristic is particularly inconvenient when semiconductor devices are considered. Intrinsic silicon has a value of k which can be 75% greater than an undoped specimen of silicon. Similar differences are exhibited by several metals which are commonly used in heat transfer equipment, e.g., aluminum, copper, and molybdenum.

Local temperature also has a marked effect on the thermal conductivity of a substance. In general, the conductivity of pure metals decreases as the temperature increases, while for alloys the reverse is true. There are, however, exceptions to this rule. For practical applications it is usually sufficiently accurate to use an average value of k when the variation in temperature is not too large.

From Eq. (11.1) it is evident that k has dimensions of heat per time, length, and temperature. A variety of units have common usage, including

SI units: W/mK
British units: Btu/hr ft°F
cgs units: cal/sec cm°C

A considerable amount of work has been done by many researchers in measuring the thermal conductivity of various solids, liquids, and gases. Table 11.1 is a brief sampling of some of these values.

11.4 ONE-DIMENSIONAL STEADY STATE CONDUCTION

A general differential equation for conduction developed from Fourier's basic law [Eq. (11.1)] is very complex and, actually, cannot be solved by exact mathematical methods. A number of simplifying assumptions are made to reduce the equation to manageable form. The simplest form is the equation for one-dimensional steady state heat conduction. This form can be used for many practical applications. When it is used, however, it must be kept in mind that the following assumptions have been made:

Heat flow has passed through the transient state and is in a steady state condition. This assumption implies

$$\frac{\partial Q}{\partial \tau} = \frac{Q}{\tau} = q$$

where q is called the heat transfer rate.

Heat flow depends on only one space coordinate. One-dimensional heat flow can be assumed with little resulting error when the lateral dimension of the heat path is large compared with the

TABLE 11.1 Values of Thermal Conductivity

Material	k W/m°C	t °C
Silver (99.999%)	428	27
Copper (99.999%)	398	27
Copper (Commercial)	331	27
Aluminum (99.996%)	237	30
Al. Alloy 2040	150	30
Silicon (Instrinsic)	144	25
Silicon (Doped*)	74	25
Window Glass	0.78	21
Oak (⊥ grain)	0.21	21
Oak (‖ grain)	0.34	21
Cork	0.043	30
Water	0.61	27
Air (1 atm)	0.027	38

Conversion Factors

[W/m °C] x (0.58) = [Btu/hr ft °F]

[W/m °C] x (2.39 x 10^{-3}) = [cal/s cm °C]

[Btu/hr ft °F] x (4.13 x 10^{-3}) = [cal/s cm °C]

*Phosphorous 1.7 x 10^{20} atoms/cm^3

length of the path, or when the thermal resistance of the lateral boundaries is large compared with the resistance of the path; two obvious examples are heat flow through a plane wall and heat flow along a wire.

The material is homogeneous so that the value of k does not depend on space coordinates, i.e.,

$$k_j \neq k_j(x,y,z)$$

The material is isotropic so that the value of k is constant, i.e.,

$$k_x = k_y = k_z = k$$

The value of k does not change with temperature, i.e.,

$$k \neq k(t)$$

When the above assumptions have been made, Eq. (11.1) reduces to

$$q = kA \frac{dt}{dx} \qquad (11.2)$$

The only variables in Eq. (11.2) are t and x, which can be separated:

$$\frac{q}{A} \int_0^{\ell} dx = -k \int_{t_{hot}}^{t_{cold}} dt$$

and integrated to give

$$\frac{q\ell}{A} = -k(t_{cold} - t_{hot})$$

It is customary to present the results in the form

$$q = \frac{kA}{\ell} \Delta t \qquad (11.3)$$

where q is the heat transfer rate, k is the thermal conductivity, A is the area normal to the heat flow, ℓ is the length of the heat flow path, and Δt is the temperature gradient $(t_{hot} - t_{cold})$.

Heat Conduction through Plane Walls

A constantly recurring condition in heat transfer analyses is the flow of heat energy through a plane wall or flat plate, as shown in Fig. 11.1. When the thickness ℓ is small compared with the lateral dimensions of the plate, Eq. (11.3) can be applied to solve for the heat flow. The same equation will determine the heat flow through a thin rod as shown in Fig. 11-1. In this case the thermal resistance of the surrounding air is much greater than that of the rod.

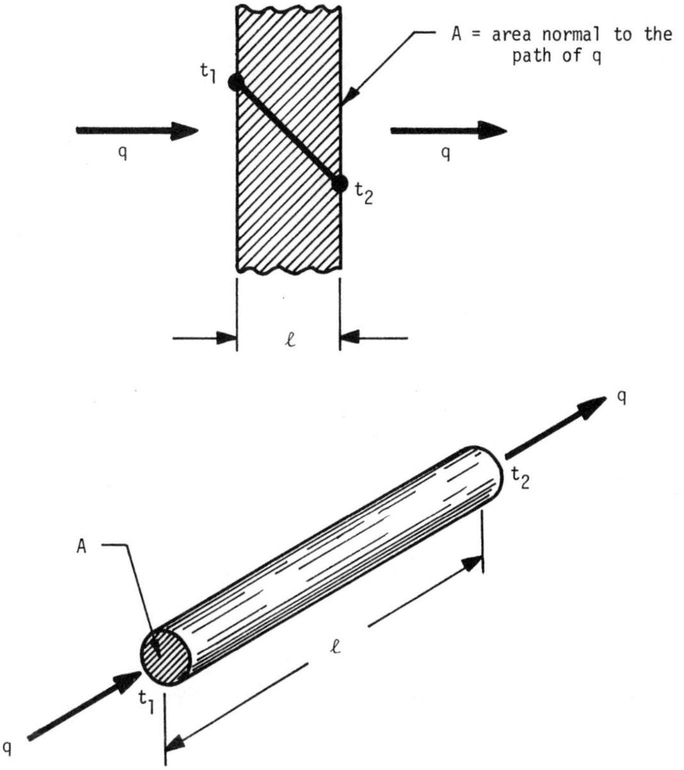

FIGURE 11.1 One-dimensional, steady state heat transfer through a plane wall and through a thin rod. Heat transfer rate q can be expressed by the one-dimensional Fourier rate equation [Eq. (11.3)].

Heat Conduction in Thick-walled Cylinders

A thick-walled cylinder has a wall thickness that is at least 10% of the diameter. If such a cylinder (Fig. 11.2) has sufficient length so that end effects are negligible, and if the inside and outside temperature t_i and t_o, respectively, are constant, the heat flow can be assumed to be one dimensional in a radial direction. The Fourier rate equation has the form

$$q = -kA \frac{dt}{dr} \tag{11.4}$$

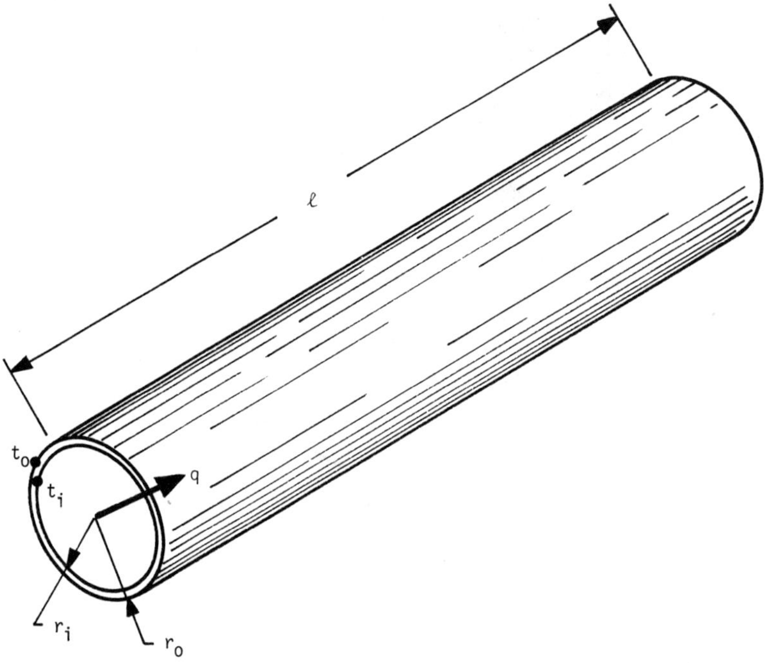

FIGURE 11.2 One-dimensional, steady state heat transfer through a thick-walled hollow cylinder.

where dt/dr is the radial temperature gradient and A is the area normal to the heat flow. The area is obviously a function of the radius

$$A = 2\pi r\ell$$

Substituting this relation in Eq. (11.4) and separating the variables:

$$\frac{q}{k2\pi\ell} \int_{r_i}^{r_o} \frac{dr}{r} = - \int_{t_i}^{t_o} dt$$

$$\frac{q}{k2\pi\ell} [\ell n\ r]_{r_i}^{r_o} = + [t]_{t_o}^{t_i}$$

$$\frac{q}{k2\pi\ell} \ell n \frac{r_o}{r_i} = t_i - t_o$$

The results are usually written in the form:

$$q = \frac{t_i - t_o}{\ell n(r_o/r_i)/2\pi k\ell} \tag{11.5}$$

Equation (11.5) indicates that the rate of heat flow in a thick-walled cylinder varies directly as the temperature difference, the cylinder length, and the thermal conductivity, and varies inversely as the natural logarithm of the ratio of the radii (or diameters).

Heat Conduction in Spherical Shells

The sphere is an interesting shape because there are no end effects as in the plane wall or cylinder. Consequently, one-dimensional heat flow equations are correct without approximations. If the interior and exterior surface temperatures are uniform and constant, and if the material is homogeneous, heat flow in a spherical shell will be radial (Fig. 11.3) and can be expressed by Eq. (11.4):

$$q = -kA\frac{dt}{dr} \tag{11.4}$$

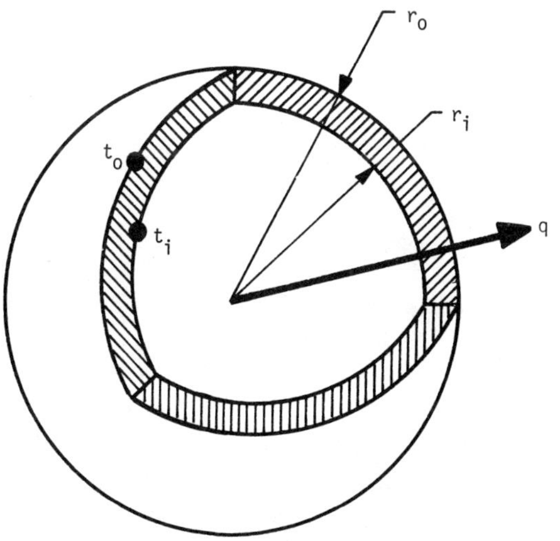

FIGURE 11.3 One-dimensional, steady state heat transfer through a spherical shell.

where dt/dr is the radial temperature gradient and the area $A = 4\pi r^2$. Substitution and separation of variables results in:

$$\frac{q}{4\pi k} \int_{r_i}^{r_o} \frac{dr}{r^2} = - \int_{t_i}^{t_o} dt$$

$$\frac{q}{4\pi k} \left[- \left[\frac{1}{r} \right]_{r_i}^{r_o} \right] = +[t]_{t_o}^{t_i}$$

$$\frac{q}{4\pi k} \left[\frac{1}{r_i} - \frac{1}{r_o} \right] = t_i - t_o$$

The results are usually written in the form

$$q = \frac{4\pi k (r_i r_o)(t_i - t_o)}{(r_o - r_i)} \tag{11.6}$$

Another useful form of Eq. (11.6) can be obtained by considering the expression for the inside and outside surface areas:

$$A_o = 4\pi r_o^2 \quad \text{and} \quad A_i = 4\pi r_i^2$$

$$\therefore \quad \sqrt{A_o A_i} = 4\pi r_i r_o$$

and substituting this relation in Eq. (11.6):

$$q = k\sqrt{A_o A_i} \left[\frac{t_i - t_o}{r_o - r_i} \right] \tag{11.7}$$

Heat Conduction in Parallelepiped Shells

A parallelepiped shell is exemplified by a furnace with a thick insulating walls as shown in Fig. 11.4. Heat flow in such a geometric shape is not one dimensional, particularly at the corners. Exact solutions are not possible and even approximate analytical solutions are cumbersome. Langmuir developed an approximate equation which gives satisfactory accuracy for practical applications:

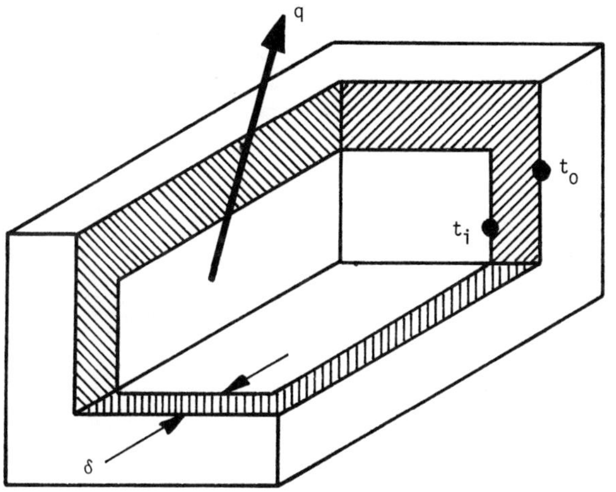

FIGURE 11.4 One-dimensional, steady state heat transfer through a parallelepiped shell.

$$q = k \frac{A_e}{\delta} (t_i - t_o) \tag{11.8}$$

In this equation δ is the thickness of the six walls, assumed to be constant, and t_i and t_o are the temperatures of the inner and outer surfaces, assumed to be constant and uniform. The value A_e is the effective area, which depends upon the relationship of the inside wall to wall lengths and the wall thickness. Representative values of A_e are shown in Table 11.2.

If the cavity of the parallelepiped shell is a cube--or approximately so--and if the walls are thick so that $A_o/A_i > 2.0$, then a modified form of Eq. (11.8) can be used:

$$q = 0.73k \sqrt{A_o A_i} \frac{t_i - t_o}{\delta} \tag{11.9}$$

Note that in Eq. (11.9) the wall thickness, δ, has replaced the difference between the radii $(r_o - r_i)$.

The equations developed for conduction heat flow in various shapes are based on a number of assumptions. In all cases, however, they give results which are satisfactory for a wide variety of heat

TABLE 11.2 Approximate Values of A_e To Be Used in Eq. (11.8)

All $L > \delta/5$	
	$A_e = A_i + 0.54\ \delta\ \Sigma L + 1.2\ \delta^2$
Two $L > \delta/5$ and one $L < \delta/5$	
	$A_e = A_i + 0.46\ \delta\ \Sigma L + 0.35\ \delta^2$
One $L > \delta/5$ and two $L < \delta/5$	
	$A_e = 6.41\ \delta\ L_{max}/\ln(A_o/A_i)$
All $L < \delta/5$	
	$A_e = 0.79\ A_i A_o$

A_i = total inside area

A_o = total outside area

L = inside wall to wall length

δ = wall thickness

transfer problems. For convenience, equations for the heat transfer rate and the thermal resistance associated with several geometric configurations are shown in Fig. 11.5.

Shape Factor for Conduction Heat Transfer

The concept of shape factor finds its greatest use in two-dimensional conduction heat transfer, but it is convenient to introduce the concept in the study of one-dimensional heat flow. The shape factor S can be defined by writing the one-dimensional Fourier rate equation in the form

$$q = kS(t_{hot} - t_{cold}) \qquad (11.10)$$

It is evident that the shape factor depends only on the geometry of a system, and can be easily identified for symmetrical shapes

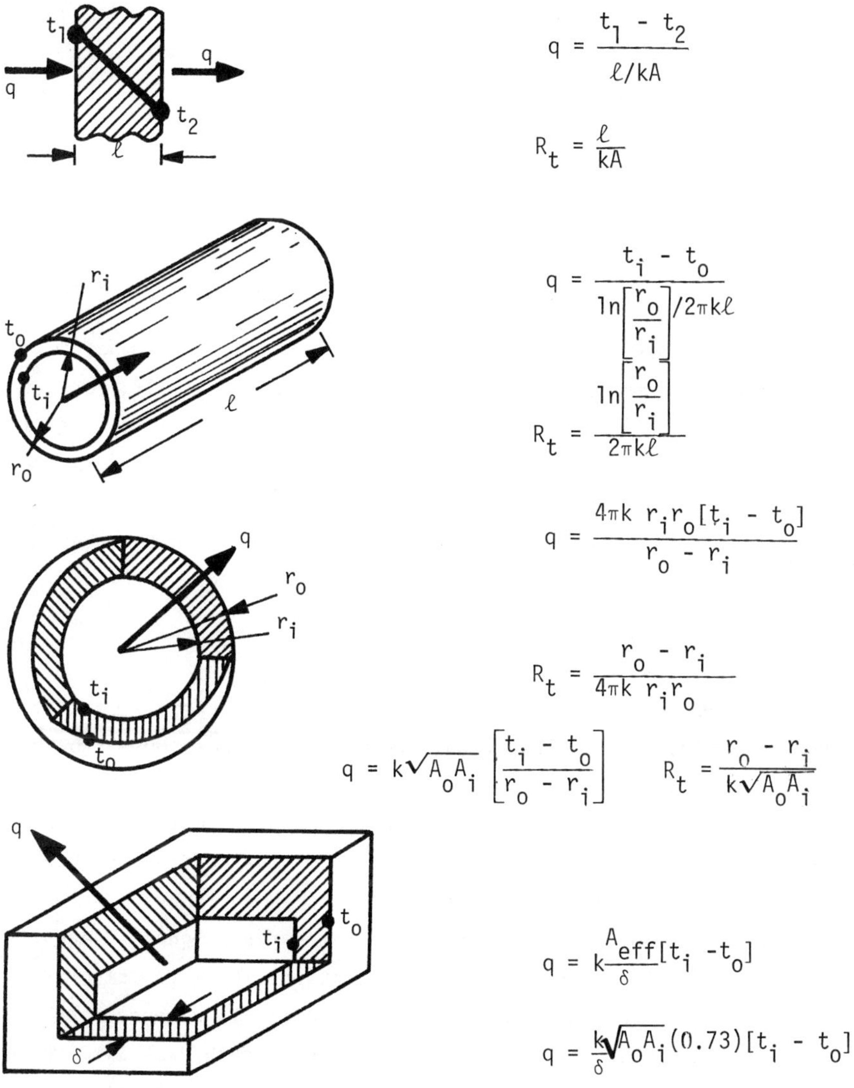

$$q = \frac{t_1 - t_2}{\ell/kA}$$

$$R_t = \frac{\ell}{kA}$$

$$q = \frac{t_i - t_o}{\ln\left[\frac{r_o}{r_i}\right]/2\pi k\ell}$$

$$R_t = \frac{\ln\left[\frac{r_o}{r_i}\right]}{2\pi k\ell}$$

$$q = \frac{4\pi k\, r_i r_o [t_i - t_o]}{r_o - r_i}$$

$$R_t = \frac{r_o - r_i}{4\pi k\, r_i r_o}$$

$$q = k\sqrt{A_o A_i}\left[\frac{t_i - t_o}{r_o - r_i}\right] \qquad R_t = \frac{r_o - r_i}{k\sqrt{A_o A_i}}$$

$$q = k\frac{A_{eff}}{\delta}[t_i - t_o]$$

$$q = \frac{k}{\delta}\sqrt{A_o A_i}(0.73)[t_i - t_o]$$

FIGURE 11.5 Heat transfer rates and thermal resistances for one-dimensional conductive heat flow in various geometries.

$$S_{wall} = \frac{A}{L}$$

$$S_{cylinder} = \frac{2\pi\ell}{\ell n(r_o/r_i)}$$

$$S_{sphere} = \frac{4\pi r_o r_i}{r_o - r_i} \tag{11.11}$$

$$S_{cube} = \frac{0.73}{\delta} \sqrt{A_o A_i}$$

Shape factor S has a dimension of length and is a number that includes all the pertinent geometric information of a body from the viewpoint of heat transfer. As a consequence, bodies having the same value of shape factor will transfer heat across a given temperature gradient in proportion to their thermal conductivities. The shape factor of a complex body can be evaluated by summing up the shape factor for each of the constituent parts of the body.

11.5 SURFACE CONDUCTANCE

When the total thermal resistance of a body is analyzed, it is necessary to consider not only internal thermal resistance, but also the thermal resistance that exists at the surface.

Suppose heat is being dissipated from a plane wall to a fluid by some convective process. The temperature distribution in the fluid can be observed by Schlieren or interference photography. The method is characterized by optical shifts which are proportional to density variations in a fluid, and consequently, proportional to temperature variations. The schematic drawing shown in Fig. 11.6 represents the type of picture which results from the process; each dark line represents a temperature gradient extending from the wall temperature t_w to the free stream temperature t_∞. In the region near the wall the gradient is very nearly linear, and this distance is designated as the film thickness δ_f. The assumption is made that the entire convective thermal resistance is concentrated in this film. The evaluation of δ_f is the subject of a field of study called boundary layer theory.

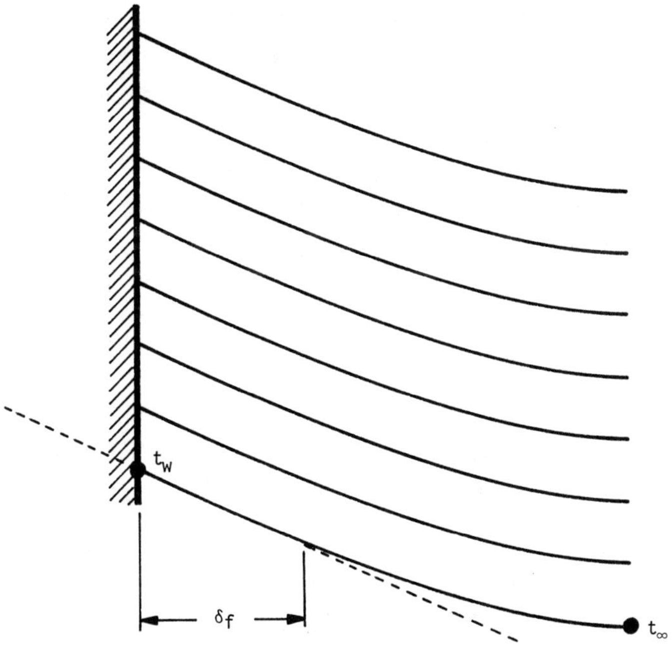

FIGURE 11.6 Thermal boundary layer: distance δ_f required for the fluid temperature to reach $0.99\ (t_w - t_\infty)$.

For the moment, let us define δ_f as the distance required for the fluid temperature to reach 99% of the total drop between t_w and t_∞. If the temperature gradient were truly uniform, and if the thermal conductivity k_f of the static film were known, the Fourier rate equation could be applied in the form

$$q = \frac{k_f}{\delta_f} A(t_w - t_\infty) \tag{11.12}$$

Because of the difficulty in determining δ_f, it is customary to let the ratio $k_f/\delta_f = h$. The heat transfer rate at the surface can then be written as

$$q = hA(t_w - t_\infty) \tag{11.13}$$

which has the familiar form of the Newton heat-transfer equation.

The term h has several different names, including *surface conductance*, *convective conductance*, and *convective heat-transfer coefficient*. A dimensional study of Eq. (11.13) shows that h has dimensions of heat per time, area, and temperature.

Particular attention should be given to area A. It actually represents the total area of the interface between the solid surface and the surrounding fluid. The area is simple when heat flows from a planar surface, but it can be more complex if heat is flowing from a finned surface.

11.6 THERMAL TRANSMITTANCE

If heat flows into one surface of a body and out of another surface of the same body, it encounters three separate and distinct thermal resistances:

Convective resistance at the hot surface (R_h)
Conductive resistance in the body (R_{body})
Convective resistance at the cold surface (R_c)

These resistances are shown in a temperature gradient sketch with a thermal circuit diagram (Fig. 11.7). The total resistance is the sum of the individual resistances in series:

$$\Sigma R = R_h + R_{body} + R_c \tag{11.14}$$

and the *thermal transmittance* U is defined as the reciprocal of the total resistance:

$$U \equiv \frac{1}{\Sigma R} \tag{11.15}$$

From the expressions for conductive and convective thermal resistance, the thermal transmittance for the plane wall shown in Fig. 11.7 can be expressed as

$$U = \frac{1}{(1/h_h) + (\ell/k)_{body} + (1/h_c)} \tag{11.16}$$

and the general expression for the heat transfer rate is

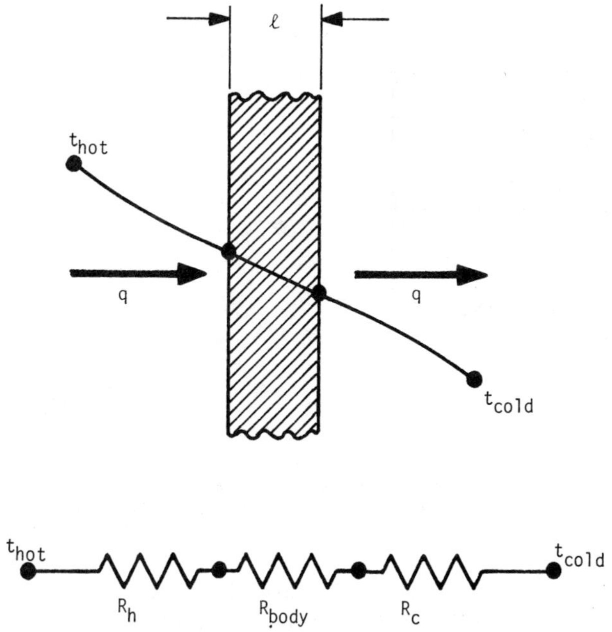

FIGURE 11.7 Thermal gradient for one-dimensional heat transfer through a plane wall and the corresponding thermal circuit diagram.

$$q = UA(t_h - t_c) \tag{11.17}$$

11.7 HEAT TRANSFER IN COMPOSITE BODIES

In many practical heat transfer systems it is common practice to cover bodies with one or more layers of material. These materials may serve a protective, insulative, or decorative function, but each one that is added contributes another thermal resistance to the system. Such systems are referred to as composite bodies and it is necessary to develop analytical techniques for heat flow through these bodies.

Composite Plane Walls

Consider heat flow through a multilayered wall as shown in Fig. 11.8. There are two convective and three conductive thermal resistances.

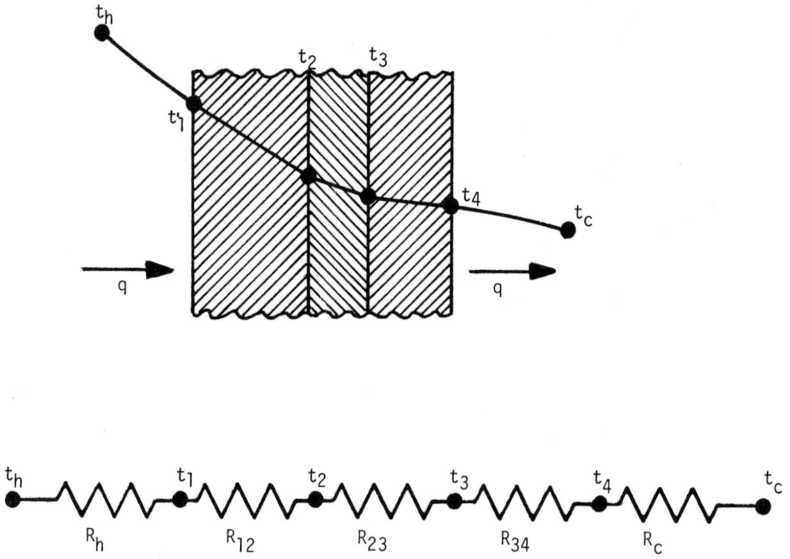

FIGURE 11.8 One-dimensional heat transfer through a multilayered plane wall and the corresponding thermal circuit diagram.

Under steady state conditions the heat transfer rate will be the same through each resistance; the value q can be determined by applying eqs. (11.3) and (11.13):

$$
\begin{aligned}
q &= [h_h A](t_h - t_i) & \therefore\ t_h - t_i &= \frac{q}{h_h A} \\
&= \left|\frac{k_{12}A}{\ell_{12}}\right| (t_1 - t_2) & \therefore\ t_1 - t_2 &= \frac{\ell_{12}}{k_{12}A} \\
&= \left|\frac{k_{23}A}{\ell_{23}}\right| (t_2 - t_3) & \therefore\ t_2 - t_3 &= \frac{\ell_{23}}{k_{23}A} \\
&= \left|\frac{k_{34}A}{\ell_{34}}\right| (t_3 - t_4) & \therefore\ t_3 - t_4 &= \frac{\ell_{34}}{k_{34}A} \\
&= [h_c A](t_4 - t_c) & \therefore\ t_4 - t_c &= \frac{q}{h_c A}
\end{aligned}
$$

A simultaneous equation for q can be obtained by adding the five equations shown above:

$$q = \frac{t_h - t_c}{(1/h_h A) + (\ell_{12}/k_{12}A) + (\ell_{23}/k_{23}A) + (\ell_{34}/k_{34}A) + (1/h_c A)} \quad (11.18)$$

This can also be expressed in terms of thermal resistances:

$$q = \frac{t_h - t_c}{R_h + R_{12} + R_{23} + R_{34} + R_c} \quad (11.19)$$

A corresponding thermal circuit diagram is shown in Fig. 11.8. The analogy also can be applied to more complex systems where thermal resistances are combined in series and parallel (Fig. 11.9):

$$q = \frac{t_h - t_c}{R_A + [R_B R_C R_D/(R_B R_C + R_C R_D + R_D R_B)]} \quad (11.20)$$

When Eq. (11.20) is applied to problems of this type the thermal conductivities of materials B, C, and D must not differ too greatly, or two-dimensional flow will result.

Composite Concentric Cylinders

A recurring problem in engineering practice has to do with a fluid at one temperature flowing through an insulated pipe whose outside surface is exposed to a fluid at another temperature. If the pipe is relatively long compared to its diameter, the heat flow will be

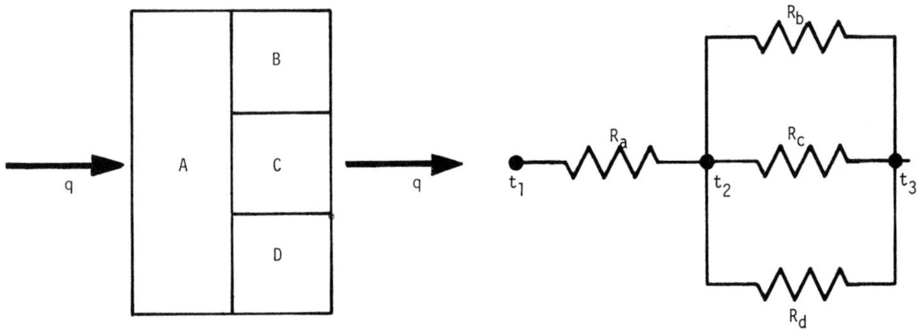

FIGURE 11.9 One-dimensional heat transfer through a composite wall with series and parallel thermal resistances.

essentially radial and the problem can be resolved into a series of concentric cylinders to which the one-dimensional heat flow equation can be applied.

Consider heat flow through two concentric cylinders as shown in Fig. 11.10. In addition to the thermal resistance of the walls, there will be a convective thermal resistance at the inner and outer surfaces. Under steady state conditions the rate of heat flow will be the same through each resistance and can be expressed by applying Eqs. (11.5) and (11.13):

$$q = 2\pi r_1 \ell h_i \,(t_i - t_1) \qquad \text{Inner surface}$$

$$= \frac{2\pi k_1 \ell}{\ell n(r_2/r_1)} (t_1 - t_2) \qquad \text{Inner cylinder}$$

$$= \frac{2\pi k_2 \ell}{\ell n(r_3/r_2)} (t_2 - t_3) \qquad \text{Outer cylinder}$$

$$= 2\pi r_3 \ell h_o \,(t_o - t_3) \qquad \text{Outer surface}$$

These four simultaneous equations can be solved by addition:

$$q = \frac{t_i - t_o}{\dfrac{1}{2\pi r_1 \ell h_i} + \dfrac{\ell n(r_2/r_1)}{2\pi k_1 \ell} + \dfrac{\ell n(r_3/r_2)}{2\pi k_2 \ell} + \dfrac{1}{2\pi r_3 \ell h_o}} \tag{11.21}$$

or expressed in the electrical analog form

$$q = \frac{(t_i - t_o)}{R_i + R_{12} + R_{23} + R_o} \tag{11.22}$$

Note that t_i and t_o refer to the free stream temperatures of the fluids at the inner and outer surfaces of the pipe, respectively.

Composite Spherical Shells

Heat flow through composite spherical shells is illustrated in Fig. 11.11 along with a thermal circuit diagram. A derivation similar to that used to develop Eq. (11.21) can be used to solve for q:

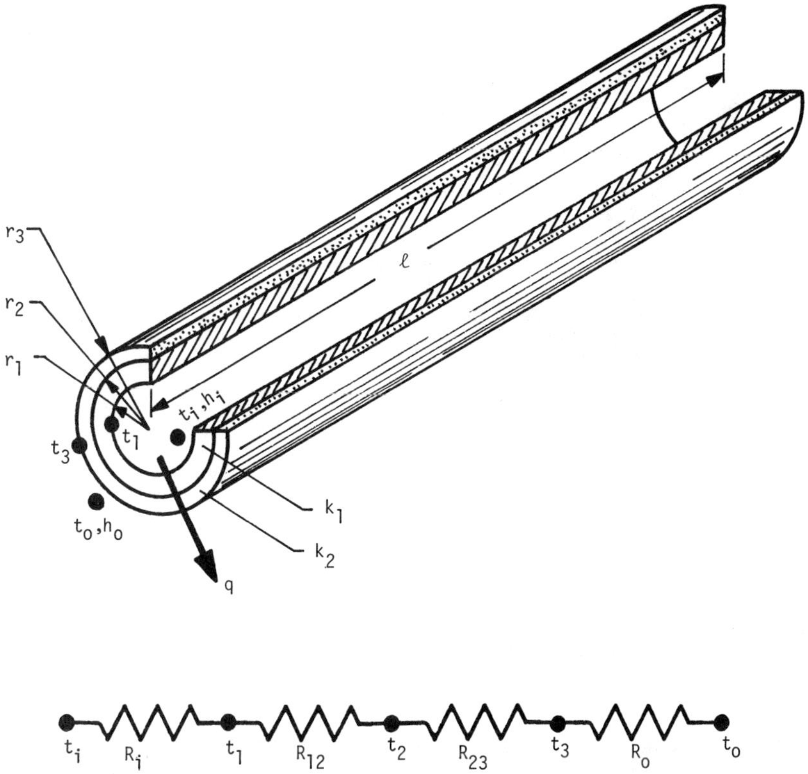

FIGURE 11.10 One-dimensional radial heat transfer through concentric cylinders and the corresponding thermal circuit diagram.

$$q = \frac{4\pi(t_i - t_o)}{(1/h_i r_1^2) + [(r_2 - r_1)/k_1 r_1 r_2] + [(r_3 - r_2)/k_2 r_2 r_3] + (1/h_o r_3^2)} \tag{11.23}$$

or, expressed in the electrical analog form,

$$q = \frac{t_i - t_o}{R_i + R_{12} + R_{23} + R_o} \tag{11.24}$$

As in the case for composite cylinders, t_i and t_o refer to the free stream temperatures of the fluids inside and outside the spheres.

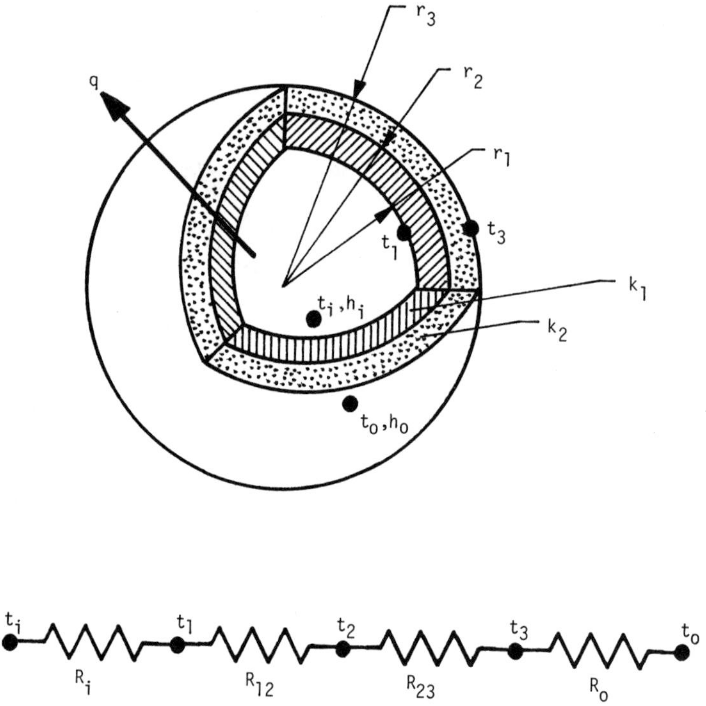

FIGURE 11.11 One-dimensional radial heat transfer through concentric spherical shells and the corresponding thermal circuit diagram.

Critical Radius (Composite Cylinders)

The addition of insulation does not always reduce the rate of heat transfer through cylindrical or spherical bodies. The reason for this is that, although the conductive resistance increases as insulation is added, the convective resistance decreases. It is evident, then, that there is a particular radius at which the thermal resistance will have a minimum value. If the outer radius of a cylinder or sphere is less than this radius (called the critical radius), then the addition of insulation will actually increase the rate of heat transfer.

In most practical cases, when metal pipes are covered with insulation, the total thermal resistance can be considered to be made up of only the conductive resistance of the insulation and the convective resistance of the outer surface. In addition, the metal pipe can be considered to be an isothermal surface, and so the temperature of the inside surface of the insulation is the same as the flowing fluid. Under these conditions the system is resolved into a single wall tube and Eq. (11.21) reduces to

$$q = \frac{2\pi k\ell(t_i - t_o)}{\ell n(r_o/r_i) + (k/h_o r_o)} \tag{11.25}$$

In this equation r_i is the inner radius of the insulation and k is the thermal conductivity. Surface conductance h_o and radius r_o refer to the outside of the insulation layer. Temperature t_i is the free stream temperature of the fluid flowing in the pipe while t_o is the free stream temperature of the fluid surrounding the insulation.

If r_i is considered fixed, then the rate of heat transfer will be a function of the outside radius and will reach a maximum when

$$\frac{dq}{dr_o} = 0$$

$$\frac{-2\pi k\ell(t_i - t_o)[(1/r_o) - (k/h_o r_o^2)]}{[\ell n(r_o/r_i) + (k/h_o r_o)]^2} = 0$$

from which

$$\frac{1}{r_o} - \frac{k}{h_o r_o^2} = 0$$

$$(r_o)\text{critical} = \frac{k}{h_o} \tag{11.26}$$

Therefore, the maximum rate of heat loss (minimum thermal resistance) will occur when the outside radius reaches the critical value. From this the following inferences can be made regarding the addition of insulation to bare pipes or wires:

If $r_{body} < r_{critical}$, then the rate of heat loss will increase as insulation is added.

If $r_{body} = r_{critical}$, then the rate of loss will be a maximum.

If $r_{body} > r_{critical}$, then the rate of heat loss will decrease as insulation is added.

Heat loss from a bare surface depends not only on convection, but also on radiation. The dependence is particularly important when natural convection occurs or when high temperatures are involved. For such cases it is necessary to use a combined surface conductance $(h_c + h_r)$. The value h_c is the convective coefficient which has been discussed in this chapter, while h_r is a radiative coefficient and will be discussed in Chap. 14. If the combined surface conductance applies to a system, the value of the critical radius for a pipe or wire becomes

$$(r_o)_{critical} = \frac{k}{(h_c + h_r)} \tag{11.27}$$

The chapter on radiation heat transfer shows that h_r is directly proportional to the emissivity of a surface. If it is desired to increase the rate of heat loss from a surface of low emissivity such as galvanized iron, it can be done by adding a thin layer of material with a high emissivity. Asbestos paper is an example of material which will accomplish this result. Electrical engineers use the principle of critical radius to obtain a combination of electrical insulation and increased cooling effects.

Critical Radius (Composite Spheres)

Using reasoning and calculations similar to those used for composite cylinders, the critical radius for composite spheres is shown to be

$$(r_o)_{critical} = \frac{2k}{h_o} \tag{11.28}$$

Conductive Resistance at an Interface

When two materials are joined together they form an interface whose thermal conductivity is often much lower than either of the materials. This is most obvious when there is dry contact, that is, two materials held together by pressure only.

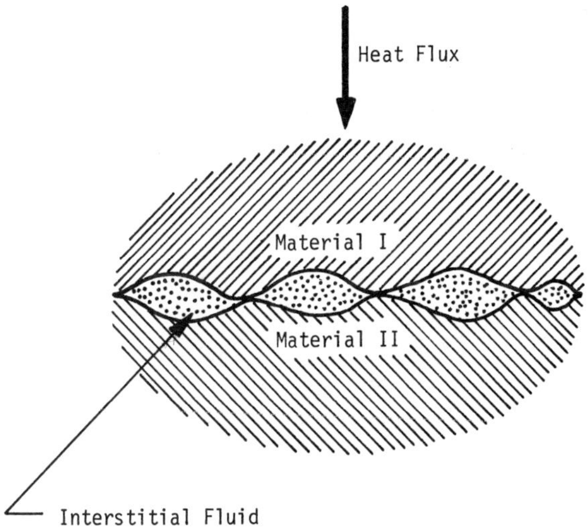

FIGURE 11.12 Contact points and interstitial fluid.

Microscopic examination shows that the surface of any material is a series of peaks and troughs so that contact with any other surface is made at only a relatively few discrete points (Fig. 11.12). As pressure is increased, some of the peaks break down, but the contact area is always less than the cross section area of either material.

A significant portion of the interface consists of a gas (usually air) trapped in the interstitial pockets between the peaks. Heat transfer through these pockets is by a combination of conduction and radiation. Since the exact geometry of an interface cannot be predicted, exact analytical heat transfer solutions are not possible. Consequently, heat transfer analysis of an interface is completely empirical.

In many heat transfer analyses the interface temperature gradient can be neglected. There are some cases, however, where a high degree of accuracy is required in analyzing temperature distribution. For these cases, a thermal contact conductance coefficient h_i is defined [4]:

$$h_i \equiv \frac{q/A}{t_1 - t_2} = \frac{q/A}{\Delta t_i} \qquad (11.29)$$

where t_1 and t_2 are the temperatures of the bounding surfaces, and q/A is the heat flow per unit area. An alternative quantity is sometimes used and referred to as the thermal contact resistance R_i:

$$R_i \equiv \frac{1}{h_i} \qquad (11.30)$$

Thermal resistance R_i can be reduced at an interface if thermal greases or indium foil are used [5]. When pressure is applied, the grease or foil flows and fills the interstitial void.

11.8 TRANSIENT HEAT TRANSFER

If the thermal environment of a body is suddenly changed, a variation in the internal temperature gradient of the body will be induced and a finite period of time will be required for the gradient to reach an equilibrium condition. The equilibrium condition is referred to as the steady state and the heat transfer process prior to the condition is called the unsteady state or transient heat transfer.

When studying unsteady state heat transfer systems, three classes of systems can be identified:

Systems in which the internal and external thermal resistances are of the same order of magnitude.

Systems in which the external thermal resistance is negligible when compared with the internal thermal resistance.

Systems in which the internal thermal resistance is negligible when compared with the external thermal resistance.

Mathematical solutions for these systems form a series of decreasing mathematical complexity with only the last one being simple enough to be described here.

There is a parameter that can be used to determine when the third system can be assumed; it is called the Biot number (Bi):

$$Bi \equiv \frac{h\delta}{k} \qquad (11.31)$$

where h is the average surface conductance of the system, δ is a characteristic length obtained from the ratio of the volume to the surface area, and k is the thermal conductivity of the body. From the definition it is evident that Bi is a dimensionless number.

A significant amount of experimental evidence indicates that if Bi < 0.1, the third system described above can be assumed and the error will be less than 5%.

The characteristic length δ has been defined as the ratio of volume to surface area. For standard geometric shapes δ can be easily determined:

Sphere $$\delta = \frac{4\pi r^3/3}{4\pi r^2} = \frac{r}{3}$$

Cylinder $$= \frac{\pi r^2 \ell}{2\pi r \ell} = \frac{r}{2}$$

Cube $$= \frac{\ell^2}{6\ell^2} = \frac{\ell}{6}$$

Newtonian Heat Transfer

The type of transient heat transfer which is described as the third system is called Newtonian heat transfer. It is approximated very closely by a body of high thermal conductivity being cooled (or heated) in air.

To derive a mathematical equation for Newtonian heat transfer, let us consider a homogeneous body of infinite thermal conductivity. The body is initially at a uniform temperature t_0 when it is suddenly immersed in a fluid at temperature t_∞. A heat balance can be made on the system for a small time interval:

> The change in thermal energy for a short time interval =
> the net heat flow from the body to the fluid for the same time interval

or expressed mathematically

$$(-c\rho V\, dt) = hA(t - t_\infty)d\tau \qquad (11.32)$$

where

c = specific heat of the body

ρ = density of the body

V = volume of the body

dt = temperature change of the body

h = average surface conductance

A = surface area of the body

t = temperature of the body

t_∞ = temperature of the fluid

$d\tau$ = time interval

The negative sign appears on the left hand side of the equation because dt is a decreasing gradient. If all the elements are considered to be constant except t and τ, Eq. (11.32) can be solved by a separation of variables:

$$\frac{dt}{t - t_\infty} = -\frac{hA}{c\rho V}\,d\tau$$

which can be integrated

$$\ell n(t - t_\infty) = -\left|\frac{hA}{c\rho V}\,\tau\right| + C_1$$

The integration constant C_1 must satisfy the initial boundary condition:

$$t = t_0 \quad \text{when} \quad \tau = 0$$

$$\therefore \quad C_1 = \ell n(t_0 - t_\infty)$$

and substituting C_1 in the solution

$$\ell n(t - t_\infty) - \ell n(t_0 - t_\infty) = -\frac{hA}{c\rho V}\,\tau$$

from which

$$\ell n\left|\frac{t - t_\infty}{t_0 - t_\infty}\right| = -\frac{hA}{c\rho V}\,\tau \qquad (11.33)$$

or

$$\frac{t - t_\infty}{t_0 - t_\infty} = e^{-(hA/c\rho V)\tau} \tag{11.34}$$

Equations (11.33) and (11.34) represent the condition where $t > t_\infty$, so that heat flows from the body to the surrounding fluid. A similar expression can be developed for the case when $t < t_\infty$, so that the body absorbs heat.

Thermal Time Constant

The quantity $c\rho V/hA$ has a dimension of time and is called the thermal time constant. When τ in Eqs. (11.33) and (11.34) equals the time constant, the temperature difference $t - t_\infty$ equals 36.8% of the initial difference $t_0 - t_\infty$. After a period of two time constants the difference drops to 13.5%, and so on. In practice a steady state condition is assumed after a period of five time constants.

Because of the thermal-electrical analogy, a thermal time constant has the same mathematical significance as an electrical time constant. However, the former is relatively long, being measured in minutes or hours, while the latter is generally a very small fraction of a second.

Graphical Representation

Equation (11.34) indicates that transient heat transfer is an exponential function. If a semilog plot is made with time τ the abcissa, a family of negative slope straight lines results. The plot, however, is dimensional in nature and does not have general applicability. A transformation to nondimensional coordinates can be made by multiplying the numerator and denominator of the exponent in Eq. (11.34) by δk:

$$\frac{t - t_\infty}{t_0 - t_\infty} = e^{-(h\delta/k)(\alpha\tau/\delta^2)} \tag{11.35}$$

where

$\mathrm{Bi} \equiv h\delta/k$ Biot number

$\mathrm{Fo} \equiv \alpha\tau/\delta^2$ Fourier number

$\alpha \equiv k/\rho c$ Thermal diffusivity

$\delta \equiv V/A$ Characteristic length

and, therefore,

$$\frac{t - t_\infty}{t_0 - t_\infty} = e^{-(Bi)(Fo)} \tag{11.36}$$

11.9 CONCLUSION

In this chapter mechanisms for heat transfer by conduction in solids, liquids, and gases are considered. The fundamental Fourier law for heat conduction is developed and formulations for one-dimensional heat conduction through plane walls, cylinders, spheres, and parallelepiped shells are presented. Concepts of thermal conductivity, shape factor, surface conductance, and thermal transmittance are defined. Heat transfer by condition through various kinds of composite bodies is discussed and methods are shown for determining the critical radius of composite cylinders and spheres.

When two metals are joined together, the resulting interface has a significant conductive resistance. This chapter includes some introductory discussion of this type of resistance. The concept of transient heat transfer is described in general terms and a formulation for one special case of one-dimensional transient heat flow is developed.

SOLVED PROBLEMS

PROBLEM 1: The inside surface of a plane wall is exposed to air at 76°F and the outside surface to air at 21°F. The inside surface conductance is 1.5 Btu/hr ft^2 °F, and the outside surface conductance is 6.5 Btu/hr ft^2 °F. If a thermocouple indicates that the inside wall temperature is 67°F, what is the outside wall temperature?

Solution: Assume steady heat flow through the wall so that the heat transfer rate on both sides has a constant value q.

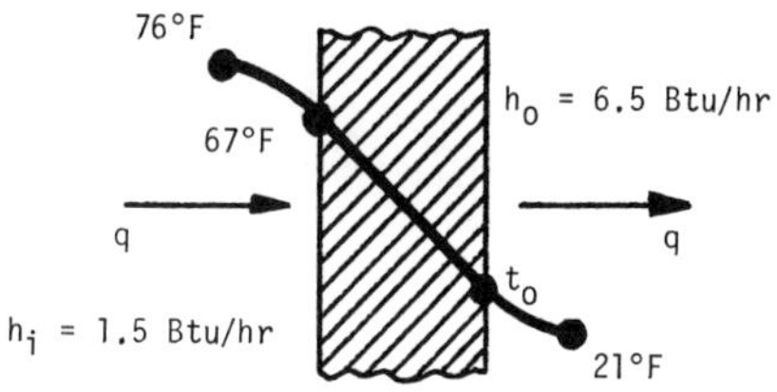

Consider one ft^2 of wall surface.

$$q = h_i A \, \Delta t_i = (1.5 \text{ Btu/hr ft}^2 \text{ °F})(1.0 \text{ ft}^2)(76 - 67)\text{°F} = 13.5 \text{ Btu/hr}$$

also,

$$q = h_o A \, \Delta t_o$$

$$13.5 = (6.5)(1.0)(t_o - 21)$$

$$\therefore \quad t_o = 23.1 \text{ °F}$$

PROBLEM 2: The inside and outside surface conductances are 2.0 and 10.0 Btu/hr ft^2 °F, respectively, and the thermal conductivity of the wall is 0.5 Btu/hr ft °F. Determine (a) the thermal transmittance and (b) the heat transfer rate for 1 ft^2 of wall surfaces.

Solution:

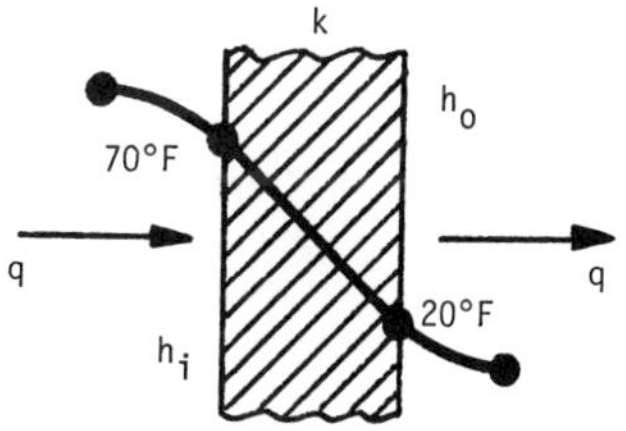

(a)

$$U = \frac{1}{(1/h_i) + (\ell/k)_{wall} + (1/h_o)}$$

where

$$h_i = 2.0 \text{ Btu/hr ft}^2 \text{ °F}$$

$$\ell = 6 \text{ in} = 0.5 \text{ ft}$$

$$k = 0.5 \text{ Btu/hr ft °F}$$

$$h_o = 10.0 \text{ Btu/hr ft}^2\ °\text{F}$$

$$U = \frac{1}{(1/2.0) + (0.5/0.5) + (1/10.0)} = 0.63 \text{ Btu/hr ft}^2\ °\text{F}$$

(b)

$$q = UA(t_i - t_o) = (0.63)(1.0)(70 - 20) = 31.5 \text{ Btu/hr}$$

PROBLEM 3: A composite wall is made up of a 1/4 in steel plate (k = 31.4 Btu/hr ft °F) and 3 in insulation (k = 0.04 Btu/hr ft °F). If the outside of the steel surface is 300°F, and the outside of the insulation is 100°F, determine (a) the heat loss and (b) the temperature at the interface of the steel and the insulation.

Solution:

(a)

$$q = \frac{t_i - t_o}{\left(\frac{\ell}{kA}\right)_{steel} + \left(\frac{\ell}{kA}\right)_{insul}}$$

and for A = 1.0 ft^2

$$q = \frac{300 - 100}{\frac{0.25 \text{ in}}{(12 \text{ in/ft})(31.4 \text{ Btu/hr ft°F})} + \frac{3.0 \text{ in}}{(12 \text{ in/ft})(0.04 \text{ Btu/hr ft °F})}}$$

$$= 32 \text{ Btu/hr}$$

(b)

$$q = \frac{t_i - t_{int}}{(\ell/k)_{steel}} = \frac{300 - t_{int}}{0.25/(12) \times (31.4)}$$

$$t_{int} = 300 - 0.02 = 299.98\ °\text{F}$$

Therefore, the temperature drop through the steel plate is negligible and the steel can be considered as an isothermal surface. Many engineering problems can be simplified by considering the conductive thermal resistance of certain materials to be zero.

PROBLEM 4: A steel pipe (k = 6.4 W/m °C) has an OD of 8.89 cm and an ID of 7.8 cm, and is covered with 1.3 cm asbestos (k = 0.19 W/m °C).

The pipe transports a fluid at 149°C and has an inner surface conductance of 227 W/m^2 °C. The outside of the pipe is exposed to air at 27°C, while the outer surface conductance is 23 W/m^2 °C. What is the heat loss of 1 m of pipe?

Solution: The conductive thermal resistance of the pipe can be neglected. Therefore, the heat loss is due only to an asbestos cylinder of 8.89 cm ID and 8.89 + 2(1.3) = 11.49 cm OD.

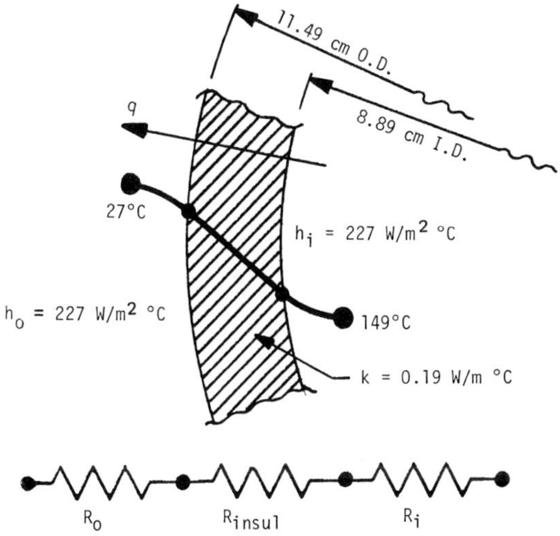

$$q = \frac{t_i - t_o}{R_o + R_{insul} + R_i}$$

where

$$t_i - t_o = 149 - 27 = 122°C$$

$$R_o = \frac{1}{D_o \pi \ell h_o} = \frac{1}{(0.1149m)(\pi)(1.0\ m)(23\ W/m^2\ °C)} = 0.12\ °C/W$$

$$R_{insul.} = \frac{\ell n(D_o/D_i)}{2\pi k \ell} = \frac{\ell n(11.49/8.89)}{2\pi(0.19\ W/m\ °C)(1.0m)} = 0.22\ °C/W$$

$$R_i = \frac{1}{D_i \pi \ell h_i} = \frac{1}{(0.0889)(\pi)(1.0)(227)} = 0.016\ °C/W$$

$$q = \frac{122}{0.12 + 0.22 + 0.016} = 343\ W$$

PROBLEM 5: The working chamber of an electrically heated furnace is a cube 0.2 m on each side and the walls are 0.15 m thick. Interior wall temperatures are to be maintained at 1100°C while the outside wall temperatures are at 150°C. If the thermal conductivity of the furnace material is 0.35 W/m °C, estimate the power consumption.

Solution: Under steady state conditions, the power consumption equals the rate of heat loss.

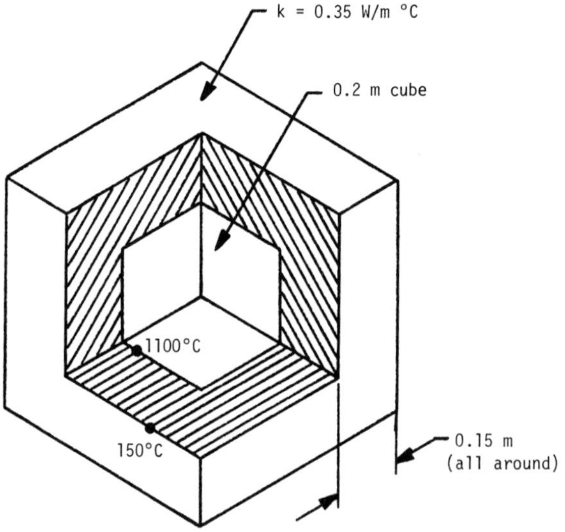

$$A_i = \text{total inside area} = 6(0.2)^2 = 0.24\ m^2$$

$$A_o = \text{total outside area} = 6(0.5)^2 = 1.5\ m^2$$

$$A_o/A_i = 1.5/0.24 = 6.25 > 2.0,$$

and since the cavity is a cube, Eq. (11.9) can be used to estimate the heat loss:

$$q = 0.73\ k\sqrt{A_o A_i}\left(\frac{t_i - t_o}{\delta}\right)$$

$$q = (0.73)(0.35)\ \sqrt{(1.5)(0.24)}\ \left(\frac{1100 - 150}{0.15}\right) = 971\ W$$

PROBLEM 6: A copper tube, 0.6 cm OD, carries hot water between two tanks; the outside surface conductance is 12 W/m^2 °C. If it is important to minimize the heat loss, should the tube be covered with an insulation whose thermal conductivity is 0.19 W/m °C?

Solution:

$$\text{Critical radius} = \frac{k}{h} = \frac{0.19\ \text{W/m °C}}{12.0\ \text{W/m}^2\ \text{°C}} = 0.016\ \text{m}$$

$$\text{radius of tube} = 0.3\ \text{cm} = 0.003\ \text{m} < 0.016\ \text{m}$$

Therefore, since $r_{tube} < r_{crit}$, the rate of heat loss will increase if the insulation is added.

PROBLEM 7: A small casting initially at 16°C is placed in a furnace at 1200°C. The ratio of volume to surface area is 0.15 m and the outer surface conductance is 85 W/m^2 °C. Thermal conductivity for the casting is 225 W/m °C and the thermal diffusivity is 0.34 m^2/hr. How much time is needed for the casting to be heated to 510°C?

Solution:

$$Bi \equiv \frac{h\delta}{k}$$

where

Bi = Biot number

h = 85 W/m^2 °C

δ = V/A = 0.15 m

k = 225 W/m °C

$$Bi = \frac{(85)(0.15)}{225} = 0.057$$

Since $Bi \ll 0.1$, the problem is one of Newtonian heating:

$$\frac{t - t_\infty}{t_o - t_\infty} = e^{-BiFo}$$

where

t = 510°C (temp. at time τ)

t_∞ = 1200°C (ambient temp.)

t_o = 16°C (temp. at $\tau = 0$)

Bi = Biot number

Fo = Fourier number

$$\frac{510 - 1200}{16 - 122} = e^{-BiFo}$$

$$e^{BiFo} = 1.71$$

$$BiFo = \ell n\ 1.71 = 0.54$$

$$\text{but } Bi \equiv \frac{h\delta}{k} \quad \text{and} \quad Fo \equiv \frac{\alpha\tau}{\delta^2}$$

$$BiFo = \left(\frac{h\delta}{k}\right)\left(\frac{\alpha\tau}{\delta^2}\right) = \frac{h\alpha}{k\delta}\tau$$

$$\tau = \frac{(BiFo)k\delta}{h\alpha} = \frac{(0.54)(225)(0.15)}{(85)(0.34)} = 0.63 \text{ hr}$$

EXERCISES

1. What is a fundamental requirement for heat to flow through a piece of material? What law of thermodynamics does this phenomenon demonstrate?
2. What is the significance of the negative sign in the Fourier rate equation?
3. A glass window pane, 2 mm thick, has a temperature of 5.07°C on one side and 4.93°C on the other. How much heat is transferred per unit area?
4. A commercial copper pipe has an i.d. of 0.9 in and an o.d. of 1.0 in. The inside of the pipe is maintained at 212°F while the outside is at 80°F. Determine the heat transfer rate per foot of pipe.
5. If the wall in solved problem 1 is 1 ft thick, determine (a) the thermal conductivity of the wall and (b) the thermal transmittance for the system.
6. In solved problem 2, determine the inside wall temperature.
7. Develop an expression for the thermal transmittance associated with a cylinder, based on the outer surface area.
8. A composite wall is made up of a 1/4 in steel plate (k = 31.4 Btu/hr ft °F), 3 in of magnesia insulation (k = 0.04 Btu/hr ft °F), and 1 in of asbestos felt (k = 0.047 Btu/hr ft °F). If the

outside of the steel is 300°F and the outside of the felt is 100°F, determine the heat loss rate.

9. A copper pipe, 3.5-in o.d. and 3.24-in i.d., is insulated with a material of k = 0.11 Btu/hr ft °F. The pipe carries a fluid at 300°F and has an inside surface conductance of 40 Btu/hr ft^2 °F. The air surrounding the pipe is at 80°F and the outer surface conductance is 4 Btu/hr ft^2 °F. What is the heat loss for 10 ft of pipe?
10. What is the heat loss for the pipe in problem 9 if the insulation is left off? How would you decide whether or not it was worthwhile to insulate the pipe?
11. The working chamber of an electrically heated furnace is 6 × 8 × 12 in and the walls are 6 in thick. The interior walls are maintained at 2000°F while the outside walls are at 300°F. Determine the power consumption if the furnace material has a thermal conductivity of 0.2 Btu/hr ft °F. [Use Eqs. (11.8) and (11.9) and compare the solutions.]
12. A 10-in o.d. pipe transports steam between two stations, and the outside ambient surface conductance is 4 Btu/hr ft^2 °F. Should insulation with k = 0.033 Btu/hr ft °F be applied to minimize heat losses?
13. A copper wire, 1/32-in diameter at 300°F, is suddenly immersed in water at 100°F. If the ambient conductance is 15 Btu/hr ft^2 °F, determine (a) the Biot number and (b) the time required for the wire to reach 120°F. (Thermal diffusivity for copper is 4.2 ft^2/hr.)

REFERENCES

1. A. G. Guy, *Essentials of Materials Science*, McGraw-Hill, New York, 1976.
2. L. H. Van Vlack, *Elements of Materials Science and Engineering*, 3rd ed., Addison-Wesley, Reading, Mass., 1977.
3. C. O. Smith, *The Science of Engineering Materials*, Prentice-Hall, Englewood Cliffs, N.J., 1977.

4. W. M. Rohsenow and J. P. Hartnett, *Handbook of Heat Transfer*, pp. 3-14 to 3-18, McGraw-Hill, New York, 1973.

5. J. H. Seely and R. C. Chu, *Heat Transfer in Microelectronic Equipment*, Marcel Dekker, New York, 1972.

12

CONVECTION HEAT TRANSFER

12.1 INTRODUCTION

Convection heat transfer can be defined as the exchange of thermal energy between two surfaces as a consequence of a relative velocity between the surfaces. Each of the surfaces may be fluid, although in most practical engineering applications one is a fluid and the other a solid surface.

In a strict sense, convection is not a distinct mode of heat transfer but rather a combination of conduction heat transfer and mass transfer. The combination occurs so frequently, however, that convection has been classified and studied as a separate and unique mechanism for the transport of thermal energy.

In 1701 Sir Isaac Newton postulated that the rate of heat transfer per unit area between a solid surface and a fluid was proportional to the temperature difference between them.

$$q = hA\,\Delta t \tag{12.1}$$

Equation (12.1) is called the Newton rate equation, and the proportionality constant h is variously referred to as the *heat transfer coefficient, surface conductance,* or *convective conductance*. The equation is not a basic law, but a definition of h. It can, however, be used in a manner of a law, since if any three elements of the equation are known, it can be used to solve for the fourth element.

Determination of h is a complex process because it depends upon many different things such as fluid properties, type of flow, geometry of the system, and surface conditions. In order to simplify the process as much as possible, several classifications of convective heat transfer have been addressed by researchers in the field. Convection is said to be either forced or free, the nature of the fluid flow is said to be laminar or turbulent, and the flow is either external or internal.

In *free (or natural) convection*, heat flows by conduction from the surface to the fluid particles in immediate contact with the surface. The energy transfer causes the motion of the particles to increase, thereby decreasing the density of the local fluid; buoyant forces then cause the particles to move to a region of lower temperature where another energy transfer occurs by conduction. A streaming action is thereby established in which lower-energy particles replace higher-energy particles at the solid-fluid interface. For the case where the solid surface is being heated by the fluid instead of cooled, a reverse process occurs.

In *forced convection*, thermal energy is transferred from the solid to the adjacent fluid particles in the same manner as free convection. The subsequent fluid motion, however, is largely the result of a pressure differential in the fluid generated by an external source such as a pump or a fan.

When the nature of fluid flow is laminar, fluid particles follow smooth continuous paths; velocity vectors are parallel at all times, and so they do not intersect. Heat transfer takes place by molecular conduction in the fluid as well as at the solid-fluid interface. In contrast, turbulent flow is characterized by irregular motion of fluid particles. Eddies exist in the fluid so that particles are continuously being mixed and rearranged. Heat transfer by conduction is augmented by heat transfer from the eddies, which transport thermal energy back and forth across streamlines. External flow occurs when a fluid flows over the outside of a solid surface; internal flow refers to a fluid flow through conduits or passageways inside a solid.

When several different fluids are used in free and forced convection systems, a wide range of heat transfer coefficients is spanned. For example, the coefficient for a free convection air system can be as low as 5 W/m^2 °C, while for a forced convection water system it can be as high as 12 000 W/m^2 °C. Table 12.1 shows the approximate range of heat transfer coefficients for practical free and forced convection systems.

12.2 METHODS OF ANALYSIS

The most challenging objective in an analysis of convective heat transfer is the determination of the heat transfer coefficient. There are five methods available to the thermal designer for analyzing any system:

TABLE 12.1 Approximate Range of Heat Transfer Coefficients for Practical Free- and Forced-Convection Systems

Fluid	Heat Transfer System	Range of h (W/m^2 °C)
Air	Free Convection	5-30
Dielectric liquids		25-100
Water		30-200
Air	Forced Convection	25-300
Dielectric liquids		50-5000
Oil		50-2000
Water		300-12000

Unit Conversion

[W/m^2 °C] x (0.1761) = [Btu/hr ft^2 °F]

[W/m^2 °C] x (2.388×10^{-5}) = [cal/s cm^2 °C]

[Btu/hr ft^2 °F] x (1.355×10^{-4}) = [cal/s cm^2 °C]

Solution of differential equations of the boundary layer
Solution of integral equations of the boundary layer
Analogy between heat, mass, and momentum transfer
Dimensional analysis
Direct measurements of iconic models

Mathematical Solutions

The first three methods listed above can be grouped under a general heading of mathematical solutions, either exact or approximate. Details of the solutions are somewhat complicated and cannot always be applied to practical heat transfer systems. The range of problems that can be solved has been increased by developments in digital computers, although the availability of computers also has drawbacks. Writers of computer programs sometimes adopt the ludicrous practice of calculating thermal and hydrodynamic quantities to several decimal places; also, the routine use of computer-aided analysis can obscure the physical significance of a problem.

In the first of the mathematical methods [1], differential equations are developed for continuity, mass diffusion, momentum, and energy, all in the boundary layer. Simplifying assumptions are then made and boundary conditions applied so that the equations can be used for some particular solution. Sometimes the calculation procedure becomes tedious and it is worthwhile to resort to approximate integral methods. In these methods momentum and energy equations are not limited to differential elements, but are integrated over the entire boundary layer. The method is simpler and can be applied more readily to turbulent flow.

The last of the mathematical methods, the analogic method, is particularly useful for analyzing turbulent processes. The earliest technique of this type was the Reynolds analogy in which he postulated that the equations for turbulent flow could be written in the same form as for laminar flow. The essence of the analogy is that the turbulent diffusivities for momentum and heat are equal at any particular point in the flow. Reynolds also assumed that the molecular heat and momentum transfer rates were negligible compared with those resulting from turbulence. This analogy is reasonably accurate when the Prandtl number is not too far from unity; this is consistent with the coincident development of the velocity and temperature profiles when Pr = 1.0.

Prandtl improved the Reynolds analogy by suggesting that two distinct regions exist in the boundary layer. In one of these, the laminar sublayer, the transfer mechanism is molecular in nature, while in the other, the Reynolds analogy still applies.

The analogy suggested by Prandtl is still an oversimplification of what takes place in the boundary layer. Many investigators have made refinements in the analogy methods to make them more adaptable to various operating conditions. In most instances a three-region boundary layer is assumed: laminar sublayer, buffer layer, and turbulent core.

Dimensional Analysis

Dimensional analysis [2,3] is a semiempirical technique for developing equations that relate measurable physical quantities by consideration of the dimensions of the quantities involved. Constants in the resulting equations must be determined by experimentation. Prior knowledge of the physical quantities that affect a process must be available because the method does not identify these quantities, it merely indicates how they affect the process.

Equations obtained by dimensional analysis are made up of experimentally determined constants and dimensionless numbers which have physical significance. Some of the dimensionless numbers used in heat transfer analysis are identified in Table 12.2.

Measurement of Iconic Models

In some cases analysis by mathematical equations or dimensional analysis may be too difficult or too inconvenient; direct measurements on iconic models will then be the best method. An iconic model is one that looks like the system it represents and it may or may not be made from

TABLE 12.2 Dimensionless Numbers Used to Correlate Convective Heat Transfer Systems

Number	Symbol	Definition	Physical significance
Nusselt	Nu	$\frac{h\ D}{k_{(fluid)}}$	$\frac{\text{total heat transfer}}{\text{conductive heat transfer}}$
Reynolds	Re	$\frac{\rho\ v\ D}{\mu}$	$\frac{\text{inertia force}}{\text{viscous force}}$
Prandtl	Pr	$\frac{c\ \mu}{k}$	$\frac{\text{momentum diffusivity}}{\text{thermal diffusivity}}$
Grashof	Gr	$\frac{D^3\ \rho^2\ g\ \beta\ \Delta t}{\mu^2}$	$\frac{\text{buoyancy force}}{\text{viscous force}}$
Stanton	St	$\frac{Nu}{Re\ Pr} = \frac{h}{v\ c\ \rho}$	$\frac{\text{heat transferred}}{\text{thermal capacity of fluid}}$
Peclet	Pe	$Re\ Pr = \frac{D\ v\ \rho\ c}{k} = \frac{D\ v}{\alpha}$	$\frac{\text{bulk heat transfer}}{\text{conductive heat transfer}}$
Rayleigh	Ra	$Gr\ Pr = \frac{D^3\ \rho^2\ g\ \beta\ c\ \Delta t}{\mu k}$	
Schmidt	Sc	$\frac{\mu}{\rho\ D}$	$\frac{\text{kinematic viscosity}}{\text{molecular diffusivity}}$
Biot	Bi	$\frac{h\ D}{k_{(solid)}}$	$\frac{\text{fluid conductance}}{\text{solid conductance}}$

the same materials. As a matter of fact, one of the advantages of iconic models is that they can be made from readily available and inexpensive materials such as wood, plastic, or cardboard. The model can be instrumented with various types of transducers and measurements of temperature, pressure, fluid velocity, and force can be made. Visual techniques can also be used with iconic models; these techniques make use of fluid dyes, smoke, and thermal paints.

Results obtained with iconic models can be used to modify mathematical solutions and to provide the constants needed in dimensional analysis. Actually, the most effective analysis can be made when the analyst is able to combine mathematical solutions, dimensional analysis, and the use of iconic models.

12.3 VARIATION IN FLUID PROPERTIES

Applications of equations from either mathematical or dimensional analysis involve the evaluation of several fluid properties. Since temperature and pressure gradients exist in heat transfer systems, the question arises as to what conditions should be assumed when fluid properties are evaluated. The problem is complicated by the fact that different fluids behave differently with changes in temperature and pressure. For example, liquid properties, with the exception of viscosity, are relatively insensitive to temperature changes. For gases, however, significant changes occur in all thermophysical properties, but the composite effect on the heat transfer coefficient is small. Liquids are generally considered to be incompressible, whereas the density of gases varies directly with pressure.

In many practical cases, changes in temperature and pressure are not large. As a result, the temperature at which properties are evaluated can be taken as an average local temperature. When the fluid flow is external in nature (see Fig. 4.6, Chap. 4), an average of the wall temperature (t_w) and the free stream temperature (t_∞) is used and designated as the film temperature (t_f):

$$t_f \equiv \frac{t_w + t_\infty}{2} \tag{12.2}$$

For internal flow (see Fig. 4.7, Chap. 4) the bulk temperature (t_b) is used as a reference temperature. Other names are the mixed mean temperature and the mixing cup temperature. It is the temperature that the fluid would assume if it were suddenly stopped, placed in an insulated container, and allowed to reach equilibrium. Mathematically it can be expressed as a function of the cross-sectional area of the duct, the local temperature and velocity profile, and the mean velocity (v_m). The mean velocity is defined such that the mass flow rate equals $\rho A v_m$:

$$t_b \equiv \frac{1}{Av_m} \int vt \, dA \tag{12.3}$$

In practice, t_b can be assumed to be the average of several temperature measurements across the duct. If heat is being transferred between a liquid and a wall, and the difference in temperature between t_w and t_b is greater than 10°F, good practice dictates that all properties should be evaluated at t_b, except for viscosity, which should be evaluated at t_w.

In most instances, changes in pressure are not sufficient to affect fluid properties, and the properties can be evaluated at the ambient pressure for the system environment. If a system is to be used at different altitudes, provisions must be made in the thermal design for variations in air density.

12.4 FREE CONVECTION

When a fluid contacts a surface that is at a higher temperature, the fluid expands and its local density is thereby decreased. The expanded fluid flows into the main body of the fluid, taking some of the heat from the hot surface with it. The resulting cooling action is called free or natural convection. When the temperature gradient is reversed, that is, the fluid is hotter than the surface, then the surface will be heated by the fluid.

When the fluid is a quiescent liquid, free convection is the primary mode of heat transfer; in the case of a quiescent gas, however, a significant amount of heat is transferred by conduction and radiation.

Free convection is the simplest of all heat transfer methods, but it also has the lowest heat-transfer coefficient. Consequently, it is useful when required heat or cooling rates are relatively small, or when the desired temperature control is not critical. A typical application of free convection cooling occurs in the design of electrical systems, particularly those using vacuum tubes.

Analyses for free convection systems are usually made empirically with the use of dimensional analysis. A generally accepted relationship states that the Nusselt number is a function of the product of Grashof and Prandtl numbers:

$$Nu = \phi(GrPr) \tag{12.4}$$

Researchers have recommended different forms of Eq. (12.4) for a variety of configurations. For example, McAdams [4] suggests that the following equations be used to determine the heat transfer coefficient for a horizontal cylinder immersed in a fluid, where the product of the Grashof and Prandtl numbers is between 10^3 and 10^9:

$$Nu = 0.53(GrPr)^{0.25} \tag{12.5}$$

and for a vertical cylinder or a vertical plane where $10^4 < GrPr < 10^9$:

$$Nu = 0.59(GrPr)^{0.25} \tag{12.6}$$

or when $10^9 < GrPr < 10^{12}$:

$$Nu = 0.13(GrPr)^{0.33} \tag{12.7}$$

Thermal properties in the last three equations are evaluated at the film temperature defined by Eq. (12.2), and the temperature difference is between the surface and the bulk fluid. The characteristic dimension used in the Grashof and Nusselt numbers can be selected from Table 12.3.

TABLE 12.3 Typical Characteristic Dimensions To Be Used in Grashof and Nusselt Numbers

Form factor	Characteristic dimension
Vertical planes	Height
Vertical cylinders	Height
Horizontal cylinders	Diameter
Horizontal square plate	Side length
Horizontal rectangular plate	Long side length
Sphere	Radius
Enclosed air spaces	Thickness of space

12.5 FORCED CONVECTION

Heat transfer coefficients that can be obtained from free convection systems are too low for many engineering designs. Consequently, a variety of fluid-moving devices are often used to establish a forced convection mode of heat transfer. The most convenient method of analyzing forced convection systems is by means of dimensional analysis. For example, the Nusselt equation states that the Nusselt number is some function of the product of powers of the Reynolds and Prandtl numbers:

$$Nu = CRe^{m}Pr^{n} \tag{12.8}$$

Numerous variations of Eq. (12.8) have been proposed [4-6] for a wide spectrum of forced convection systems. In considering several examples of these variations, it is necessary to differentiate between internal and external flow systems.

Internal Flow Forced Convection

The problem of determining the heat transfer rate for a fluid flowing through a conduit occurs frequently in practice. The most common examples are heat exchangers used in power, heating, and refrigerating cycles.

A widely used empirical relation for fully developed turbulent flow through smooth tubes has the form

$$Nu = 0.023Re^{0.8}\ Pr^{n} \tag{12.9}$$

Properties in Eq. (12.9) are evaluated at the mixing cup temperature [Eq. (12.3)], and the characteristic dimension is the hydraulic diameter [Eq. (4.4)]. Exponent n has different values depending on whether the fluid is being heated or cooled:

$$n = \begin{cases} 0.4 & \text{heated} \\ 0.3 & \text{cooled} \end{cases}$$

External Flow Forced Convection

The existence of a thermal boundary layer was introduced in Chap. 4. When heat is transferred between a fluid and a solid surface, the thermal resistance is assumed to be concentrated in the boundary layer.

A simple case of external flow occurs when fluid flows over a flat plate. For such a condition the heat transfer coefficient can be found by applying Eq. (12.1). When fluid flows past a cylinder or sphere, however, the condition is more complex because the fluid separates from the surface at a point very close to the leading edge. Beyond the point of separation, turbulent flow occurs and turbulent eddies can be observed.

Much research has been done concerning air flow past single cylinders. According to McAdams [4] all the data collected can be correlated by an equation of the form

$$Nu = CRe^{n} \tag{12.10}$$

Note that the Prandtl number does not appear in Eq. (12.10) because it varies very little for air over the range of data collected. Knudsen and Katz [7] found that it was necessary to include the one-third power of the Prandtl number in Eq. (12.10) to correlate data for liquids. Thermophysical properties in Eq. (12.10) are evaluated at the film temperature [Eq. (12.2)] and the characteristic dimension is the cylinder diameter. Constants for Eq. (12.10) are shown in Table 12.4.

12.6 AUGMENTATIVE TECHNIQUES

There are natural limitations to how much heat can be transferred in convective systems, as shown in Table 12.1. When a thermal system is being designed, other limitations are often imposed by specifications for size, weight, reliability, accessibility, and cost. As a consequence, thermal designers are always looking for techniques that will enhance or augment standard modes of heat transfer.

TABLE 12.4 Constants To Be Used with Eq. (12.10)

Re	C (gases)	C (liquids)	n
0.4 - 4	0.89	0.99 $Pr^{1/3}$	0.33
4 - 40	0.82	0.91 $Pr^{1/3}$	0.39
40 - 4 000	0.62	0.68 $Pr^{1/3}$	0.47
4 000 - 40 000	0.17	0.19 $Pr^{1/3}$	0.62
40 000 - 400 000	0.024	0.027 $Pr^{1/3}$	0.81

Bergles [8] surveyed and evaluated several basic techniques for increasing heat transfer coefficients in convective systems. He categorized them as follows:

Surface promoters
Displaced promoters
Vortex flow
Surface or fluid vibrations
Electrostatic fields
Fluid additives

Generally speaking, these techniques promote turbulence in a system over and above that caused by the convective process. The net result of the additional turbulence is an increase in the heat transfer rate.

Very often, special design techniques can be used to achieve the best possible transfer rates. Examples of these techniques were given by Kraus [9] and Seely and Chu [10].

12.7 CONCLUSION

This chapter covers the basic principles of convective heat transfer systems, both free and forced. Methods of analysis by mathematical solutions, dimensional analysis, and measurements of iconic models are discussed, and considerations are given to variations in fluid properties for internal and external flow patterns. Some formulations of non-dimensional nature are presented. Several techniques for augmenting heat transfer are described for overcoming the natural limitations of convective systems.

SOLVED PROBLEMS

PROBLEM 1: An electrically heated vertical plate, 5 in square, has a temperature of 150°F and is being cooled by natural convection in 50°F air. What is the heat transfer rate from both sides of the plate?

Solution: Thermal properties are evaluated at the film temperature t_f.

$$t_f = \frac{t_w + t_\infty}{2} = \frac{150 + 50}{2} = 100°F$$

From table A-3 [5], at 100°F:

$$\frac{g\beta\rho^2}{\mu^2} = 1.76 \times 10^6 \ (°F\ ft^3)^{-1}$$

$$Pr = 0.72$$

$$k = 0.015 \text{ Btu/hr ft °F}$$

$$Gr \equiv \left(\frac{g\beta\rho^2}{\mu^2}\right) D^3 \ \Delta T$$

where D = 5 in = 0.42 ft and ΔT = 150 - 50 = 100°F

$$Gr = (1.76 \times 10^6)(0.42)^3(100) = 1.3 \times 10^7$$

$$(GrPr) = (1.3 \times 10^7)(0.72) = 9.4 \times 10^6$$

from Eq. 157:

$$Nu = 0.59\ (GrPr)^{0.25} \quad \text{but} \quad Nu \equiv \frac{hD}{k}$$

$$\therefore \quad h = (0.59)(9.4 \times 10^6)^{0.25}\ \frac{0.015}{0.42} = 1.2 \text{ Btu/hr ft}^2\ °F$$

$q = hA\ \Delta T$ but heat is transferred from both sides

$$q = 2\left[(1.2)\left(\frac{5 \times 5}{144}\right)(100)\right] = 41.6 \text{ Btu/hr}$$

PROBLEM 2: Cooling water at an average temperature of 70°F flows through a tube 0.9 in ID, with an average velocity of 7 ft/s. What is the heat transfer coefficient when the flow is fully developed?

Solution: Thermophysical properties for water taken from table A-3 [5] at 70°F:

$$\rho = 62.3 \text{ lbm/ft}^3$$

$$\mu = 6.58 \times 10^{-4} \text{ lbm/ft s}$$

$$Pr = 6.82$$

$$k = 0.347 \text{ Btu/hr ft °F}$$

Characteristic dimension D = hydraulic diameter = 0.9 in = 0.075 ft

$$Re \equiv \frac{\rho v D}{\mu} = \frac{(62.3 \text{ lbm/ft}^3)(7 \text{ ft/s})(0.075 \text{ ft})}{(6.58 \times 10^{-4} \text{ lbm/ft s})} = 49\ 700$$

Since Re > 2300, the flow is turbulent. The expression for the Nusselt number for fully developed turbulent flow with the fluid being heated is

$$Nu = 0.023\, Re^{0.8}\, Pr^{0.4} \quad \text{but} \quad Nu \equiv \frac{hD}{k}$$

$$\therefore \quad h = (0.023)(49700)^{0.8}(6.82)^{0.4}\,\frac{0.347}{0.075} = 1311 \text{ Btu/hr ft}^2 \text{ °F}$$

PROBLEM 3: Air at 1 atm pressure and a mixing cup temperature of 450 K flows through a 3 cm diameter tube with a velocity of 6 m/s. Determine the heat transfer rate per unit length of tube if a constant heat flux condition is maintained at the tube wall and, the wall temperature is 10°C above the air temperature.

Solution: Properties of air at 450 K can be found in *Convection Boiling and Condensation* by J. G. Collier (table A-7):

$$\rho = 0.783 \text{ kg/m}^3$$

$$k = 0.0371 \text{ W/m °C}$$

$$\mu = 2.48 \times 10^{-5} \text{ Ns/m}^2 \text{ or kg/ms}$$

$$Pr = 0.683$$

Characteristic dimension D = hydraulic diameter = 0.03 m

$$Re \equiv \frac{\rho v D}{\mu} = \frac{(0.783\ kg/m^3)(6\ m/s)(0.03\ m)}{2.48 \times 10^{-5}\ kg/ms} = 5683$$

Since Re > 2300, the flow is turbulent and Eq. 110 can be used to determine the heat transfer coefficient:

$$Nu = 0.023\ Re^{0.8}\ Pr^{0.4} \quad \text{but} \quad Nu \equiv hD/k$$

$$\therefore \quad h = 0.023(5683)^{0.8}\ (0.683)^{0.4}\ \frac{0.0371\ W/m\ °C}{0.03\ m} = 24.6\ W/m^2\ °C$$

The heat transfer rate q can be expressed as

$$q = hA\ \Delta T$$

where

$$A = \pi\ D\ L$$

$$\Delta T = 10°C$$

So the heat transfer rate per unit length is

$$q/L = (24.6\ W/m^2\ °C)(\pi\ 0.03\ m)(10°C) = 23.2\ W/m$$

PROBLEM 4: Air at 1 atm pressure and 25°C flows past a horizontal 5-cm diameter cylinder with a velocity of 46 m/s. If the surface of the cylinder is kept at 135°C, determine the rate of heat flow from the cylinder.

Solution: External flow properties are evaluated at the film temperature t_f:

$$t_f \equiv \frac{t_w - t_\infty}{2} = \frac{25 + 135}{2} = 80°C\ (353\ K)$$

Thermal properties for air at 350 K can be found in *Convective Boiling and Condensation* by J. G. Collier (table A-7):

$$\rho = 0.998\ kg/m^3$$

$$k = 0.03\ W/m\ °C$$

$$\mu = 2.08 \times 10^{-5}\ kg/s\ m$$

Characteristic dimension D = diameter of cylinder = 0.05 m

$$Re \equiv \frac{\rho v D}{\mu} = \frac{(0.998\ kg/m^3)(46\ m/s)(0.05\ m)}{(2.08 \times 10^{-5}\ kg/s\ m)} = 110\ 400$$

The coefficient of heat transfer can be found by applying Eq. (12.10):

$$Nu = C\ Re^n \quad \text{and} \quad Nu \equiv hD/k$$

where from Table 12.4

$$C = 0.024$$

$$n = 0.81$$

$$h = (0.024)(110\ 400)^{0.81}\ \frac{0.03\ W/m\,°C}{0.05\ m} = 175\ W/m^2\ °C$$

The rate of heat transfer q can be found by applying the Newton rate equation:

$$q = hA\ \Delta T$$

where

$$A = \pi DL$$

$$\Delta T = 135 - 25 = 110°C$$

The heat transfer rate per unit length of cylinder is

$$q/L = (175\ W/m^2\ °C)(\pi\ 0.05\ m)(110°C) = 3024\ W/m$$

EXERCISES

1. How is the heat transfer coefficient h defined for convective systems?
2. What is the fundamental difference between free and forced convection?
3. Show that the Reynolds number is a dimensionless quantity.
4. Calculate the film temperature for the outside wall surface in solved problem 1, Chapter 11.
5. What is the heat transfer rate from the plate of solved problem 1, Chapter 11, if the radiation heat transfer coefficient is 1.5 Btu/hr Ft^2 °F?

6. A 1-in diameter metal cylinder, 1-1/2 ft long, has a surface temperature of 130°F. If it is placed vertically in a body of water at 70°F, what is the rate of heat transfer from the cylinder to the water?
7. Air at 1 atm and 400°F mixing cup temperature flows through a 1-in diameter tube with a velocity of 20 fps. Determine the heat transfer rate per unit length of tube if the tube temperature is maintained at a temperature 20°F greater than the air temperature.
8. A cylindrical fin of 2-in diameter is maintained at 310°F by a stream of atmospheric air of 90°F. The air is moving at 160 fps. What is the rate at which heat is being removed from the fin?

REFERENCES

1. W. M. Kays, *Convective Heat and Mass Transfer*, McGraw-Hill, New York, 1966.
2. P. W. Bridgman, *Dimensional Analysis*, Yale University Press, New Haven, Conn., 1963.
3. S. J. Kline, *Similitude and Approximation Theory*, McGraw-Hill, New York, 1965.
4. W. H. McAdams, *Heat Transmission*, 3rd ed., McGraw-Hill, New York, 1954.
5. F. Kreith, *Principles of Heat Transfer*, 2d ed., International Textbook, Scranton, Pa., 1967.
6. W. M. Rohsenow and P. J. Hartnett, *Handbook of Heat Transfer*, McGraw-Hill, New York, 1973.
7. J. D. Knudsen and D. L. Katz, *Fluid Dynamics and Heat Transfer*, McGraw-Hill, New York, 1958.
8. A. E. Bergles, Survey and evaluation of techniques to augment convective heat and mass transfer, in *Progress in Heat and Mass Transfer*, Vol. I, Pergamon, New York, 1969.
9. A. D. Kraus, *Cooling Electronic Equipment*, Prentice-Hall, Englewood Cliffs, N.J., 1965.
10. J. H. Seely and R. C. Chu, *Heat Transfer in Microelectronic Equipment*, Marcel Dekker, New York, 1972.

13

BOILING HEAT TRANSFER

13.1 INTRODUCTION

Boiling heat transfer is a mode of heat transfer that occurs because of vaporization. *Vaporization* is a process in which a substance is changed to a vapor state. It is a general term and includes: (1) *evaporation*, which is a conversion from a liquid to vapor state that occurs at the liquid-vapor interface only; (2) *boiling*, which is a conversion from a liquid to vapor state that takes place in the interior of the liquid through the formation of vapor bubbles; and (3) *sublimation*, which is a direct conversion of a solid to vapor state. In this chapter only the second type of vaporization is treated in detail.

In a vaporization process thermal energy is absorbed (or released) so that any substance that is vaporized (or condensed) represents a heat transfer system. Nature makes use of this condition in cooling mammals. Body fluids formed by the sweat glands are vaporized and, by means of this change in phase, the body of the mammal is kept at a constant temperature even though the temperature of the environment is higher and fluctuating. Long ago, man learned to cool quantities of water by storing it in porous jars and allowing the water to evaporate. A more efficient means for maintaining a low temperature is to place solid phase water (ice) in a compartment and allowing it to convert to the liquid phase. An amount of energy corresponding to the heat of fusion will be absorbed from the compartment and the temperature will be lowered. In another application, solid phase carbon dioxide (dry ice) is allowed to sublime in a closed space, and an amount of energy corresponding to the heat of sublimation is absorbed. Modern refrigeration techniques rely on special refrigerating liquids that are reduced to a low energy level by suitable equipment and which then absorb heat from a compartment through a vaporization process. Modern steam heating systems also make use of phase changes. Liquid phase water is converted to steam by means of a combustion process. Each gram of water absorbs an amount of heat which corresponds to the heat of vaporization (540 cal at atmospheric pressure); this thermal energy is subsequently released in the compartment to be heated by means of a condensation process.

The phase change from liquid to vapor in the latter application is accomplished by a boiling process, which is different from an evaporation process. Evaporation takes place only at the free surface of the liquid, and the thermal energy required to effect it comes from the liquid itself so that the liquid temperature is lowered during evaporation. Boiling, on the other hand, depends on the energy supplied from a submerged surface and is essentially a constant temperature process. The rate of evaporation is proportional to the liquid temperature and increases until the boiling point is reached. As a consequence, low boiling point liquids are more volatile at ordinary temperatures than those with high boiling points.

13.2 BOILING HEAT TRANSFER NOMENCLATURE

A number of terms are used to prefix the word boiling. Some of these terms have overlapping meanings and some of them are given varying interpretations by practitioners of different disciplines.

The term *nucleate* boiling has general acceptance and means that vapor bubbles are initiated in a liquid due to the presence of certain kinds of nuclei. These nuclei may be vapor-filled cavities in the heater surface, small unwetted foreign particles, noncondensable gas bubbles suspended in the liquid, or a cluster of high energy molecules that are a result of thermal fluctuations of liquid molecules.

The latter type of nucleus is characteristic of *homogeneous* or *volume* boiling, which occurs in pure liquids heated by radiation or in solution-type nuclear reactors. The first three types of nuclei are characteristic of heterogeneous boiling, which occurs in most practical applications.

Homogeneous boiling is relatively rare because heterogeneous boiling occurs long before the relatively large superheats which would be needed for homogeneous boiling are reached. Homogeneous boiling seldom is observed in water and never in practical engineering systems using water; it does, however, occur in some organic liquids.

Two major classifications of boiling are pool and flow boiling. *Pool* or *pot* boiling takes place when a liquid is confined in a container and a heater is submerged in the liquid; movement of the liquid is due only to changes in density which cause natural convection currents. *Flow* or *forced convection* boiling describes a boiling system in which a liquid is pumped through a duct between plates or over a matrix of submerged heat exchanger tubes.

When the temperature of the bulk of liquid is below saturation, the type of boiling is referred to as *subcooled* or *local* boiling. If the liquid temperature is at, or above, saturation the term *saturated* or *net* boiling is used.

Film boiling describes a condition where a film of vapor forms between the heated surface and the liquid being boiled. Some authors use the term *incipient* boiling to designate the early stages of nucleate boiling, when vapor bubbles just begin to form. The prefix *bulk* is sometimes used to designate the advanced stages of nucleate boiling in a saturated pool; the term bulk boiling is also used as a synonym for homogeneous or volume boiling.

Another method of classifying boiling information is by the type of heating that causes the boiling; two classifications are constant heat input and constant temperature. A *constant heat input* system is typified by an electric wire or a nuclear reactor, while a *constant temperature* system is typified by a condensing vapor, such as steam flowing through a duct.

13.3 BOILING REGIMES

There are arguments in favor of considering boiling heat transfer to be a special case of convective heat transfer, either natural or forced. Consider that the rate of convection heat transfer q for a system is expressed analytically by the Newton rate equation:

$$q = hA\ \Delta t \tag{12.1}$$

where h is the heat transfer coefficient, A is the area involved in the heat transfer, and Δt is some well-defined temperature difference. When heat is being dissipated from a solid surface to a fluid, it is customary to let Δt be the difference between the surface temperature and the temperature at the limit of the thermal boundary layer. An analogous equation is used for boiling heat transfer:

$$q'' = h_b\ \Delta t \tag{13.1}$$

where $q'' = q/A$ is called heat flux and h_b is the boiling heat transfer coefficient. Several different values of Δt are used in the analysis of boiling heat transfer. A widely used value is called "wall superheat" or "surface temperature excess" Δt_{sat}, which is defined as the difference between the wall temperature of the heating surface t_w and the saturation temperature of the liquid t_{sat}:

$$\Delta t_{sat} \equiv t_w - t_{sat} \tag{13.2}$$

The value of h_b for a boiling system changes as the system passes through different regimes. These regimes can be shown on a boiling curve.

Boiling Curve

The boiling curve is generally plotted as log-log coordinates, with heat flux q" versus wall superheat Δt_{sat}. In order to obtain data for plotting a boiling curve, let us imagine an experiment in pool

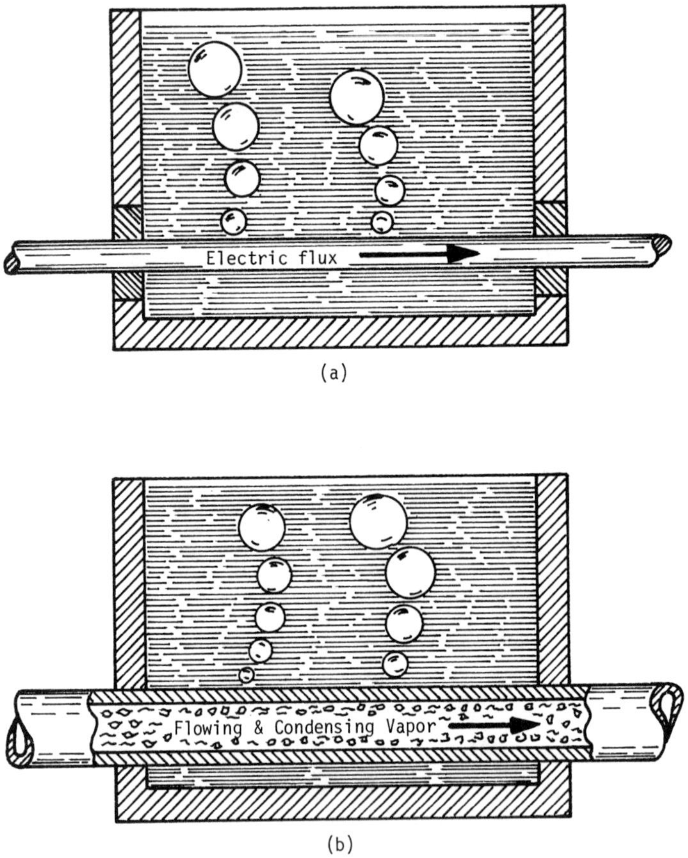

FIGURE 13.1 Pool boiling models for developing a boiling curve: (a) electric resistance heat for constant energy input, and (b) condensing vapor heating for constant temperature surface.

boiling as illustrated in Fig. 13.1. The liquid to be boiled is in a large open container, and two different methods of heating are used to provide heat flux to the liquid.

In Fig. 13.1a an electric resistance heater provides constant energy input, while in Fig. 13.1b heating is accomplished by passage of a condensing vapor to provide a constant surface temperature system.

Initially, as q'' and Δt_{sat} are increased, the liquid is heated by convection. It might be thought that, when the liquid close to the heater surface reaches its saturation temperature, vapor bubbles will form on the heater surface. However, careful measurements and observations in many experiments [1] show that this condition does not occur. The liquid must be superheated to some degree before bubbles form. As the liquid is heated it becomes less dense and rises toward the free surface where vaporization takes place. Therefore, the cooling process on the heater surface is one of natural convection. On the boiling curve of Fig. 13.2 the natural convection region from 1-2 appears as a straight line on a log-log plot.

At point 2 the liquid superheat has increased to a point where vapor bubbles begin to form; this point on the curve is referred to

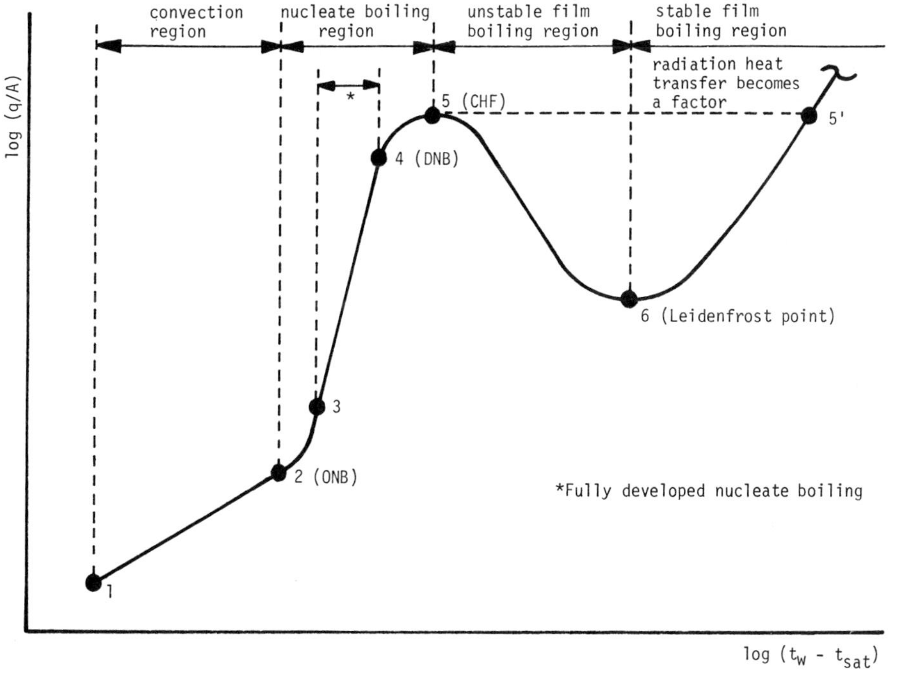

FIGURE 13.2 Boiling curve showing regimes of boiling heat transfer (not to scale).

as the "onset of nucleate boiling" (ONB). Careful observation of the heating surface will show that the bubbles appear at a few preferred locations. As the bubbles detach from the heated surface, they form a column of bubbles moving toward the free surface of the liquid. Line 2-3 is the "knee" of the boiling curve and represents a region of changing heat transfer coefficient.

When point 3 is reached, a sufficient number of nucleation sites will have been activated to establish fully developed nucleate boiling. The line 3-4 representing this region will once again be straight, but its slope will be much greater than in the convection region. This means that very large heat transfer rates are possible with relatively small temperature driving forces; the characteristically high heat-transfer coefficient h_b of nucleate boiling is obtained.

As heat flux is increased further, more nucleation centers are activated and a large amount of vapor forms on the heater surface. This causes the heat transfer coefficient to decrease from point 4 to 5. Point 4 is known as the "departure from nucleate boiling" (DNB).

At some critical value of heat flux at point 5, a blanket of vapor forms over the entire surface of the heater and heat transfer is severely curtailed. The vapor film is unstable and it forms, collapses, and reforms repeatedly. Therefore, region 5-6 is referred to as one of "unstable film boiling." If the electrical heater of Fig. 13.1 is being used when point 5 is reached, the process cannot remove an amount of heat equal to the constant electrical energy input. The resulting discrepancy causes a rise of internal energy in the heater, with a decrease in heat flux. The system is unstable and the value of Δt_{sat} will spontaneously jump to point 5'. If data points for line 5-6 are to be obtained, it is necessary to use the condensing vapor heat of Fig. 13.1; the system will then be one of constant surface temperature.

Wall temperatures represented by point 5' are well above the melting points of many metals. Consequently, point 5 is sometimes referred to as the "burnout point." Other names that appear in the literature for this point are "peak heat flux" and "critical heat flux"; some authors refer to the condition as "boiling crisis."

In the range of unstable film boiling, some wetting of the heater surface by the liquid can occur; this accounts for the instability of the vapor film. After a point such as 6 has been reached, the wall temperature is so high that a steam cushion appears between all liquid droplets and the heater wall, and no wetting is possible. The region after point 6 is, therefore, one of "stable film boiling." Point 6 is referred to as the "spheroidal state" or the "Liedenfrost point." Wall temperatures in the stable film boiling range are at a high enough level that radiation heat transfer becomes significant, resulting in increased values of boiling heat transfer coefficients.

Variations in the Boiling Curve

The boiling curve of Fig. 13.2 is representative of all types of boiling systems. The actual coordinates of heat flux and temperature difference obviously depend on the nature of the system. For example, flow boiling in a tube is initially characterized by forced convection cooling. Other factors such as the degree of subcooling or superheating will cause variations in the boiling curve. Operating pressures, surface roughness, amount of noncondensables, and other parameters also influence the shape of the curve. These parameters are the essence of boiling heat transfer research and will be addressed in subsequent sections.

13.4 HEAT TRANSFER MECHANISMS IN NUCLEATE BOILING

The relative importance of several mechanisms of heat transfer in nucleate boiling have been debated by a number of analysts in the field. Several models have been proposed [2-4]. It appears that heat is transferred in a variety of ways during a nucleate boiling process. The best that can be said is that the high heat-transfer rates associated with nucleate boiling are due to the influence of several mechanisms.

Latent Heat Transfer

About 2257 joules of heat are absorbed for each gram of water that is vaporized at 1 atm pressure. A number of studies have shown that this mechanism is responsible for only a small fraction of the total heat transfer of a boiling system. For example, Gunther and Kreith [5] showed that latent heat transport accounted for about 2% of the total heat flux when water was boiled in a subcooled pool at 1 atm.

Microconvection in the Boundary Layer

The most widely accepted mechamism of heat transfer in nucleate boiling is one in which a pumping action is established by the growth and collapse of vapor bubbles. Figure 13.3 is a schematic diagram of the mechanism.

A bubble is initiated on a hot surface in a superheated layer of liquid. As the bubble grows to some volume, it forces an equal volume of superheated liquid out of the thermal boundary layer into the main body of liquid; the main body of liquid is at some lower temperature. If the liquid is subcooled, the bubble will collapse on the surface; if the liquid is above the saturation temperature, the bubble will detach and grow as it rises. In either case the region originally occupied by the bubble becomes, momentarily, a vacuum. The vacuum is immediately filled with liquid moving with a relatively high velocity and at a temperature considerably lower than the wall temperature. This violent agitation adjacent to the heated surface is called "microconvection in the boundary layer" and is responsible for a large amount of heat transfer in a nucleate boiling system.

Vaporization of a Microlayer of Liquid

Another mechanism that is proposed assumes that a microlayer of liquid forms under a growing bubble due to viscosity effects (Fig. 13.4). The presence of the microlayer is attributed to the retarding action of viscous forces on the surface. The liquid in this layer is rapidly vaporized, causing a rapid removal of heat from the surface, with a corresponding temperature reduction.

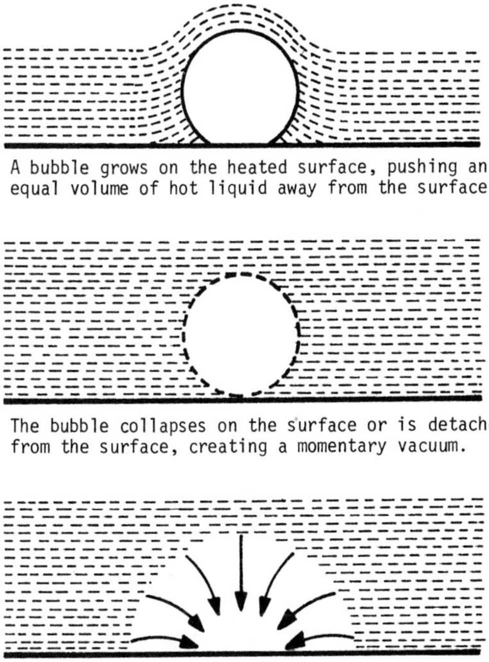

(a) A bubble grows on the heated surface, pushing an equal volume of hot liquid away from the surface.

(b) The bubble collapses on the surface or is detached from the surface, creating a momentary vacuum.

(c) Liquid moving with a high velocity and at a temperature less than that of the surface rushes into the evacuated region and is sprayed on the surface.

FIGURE 13.3 Schematic drawing of the boiling heat transfer mechanism called "microconvection in the boundary layer."

Thermocapillarity (Jetting)

Photographic analysis of boiling from a heated surface in a highly subcooled liquid appears to show jets of hot liquid being flung outward from the bubbles (Fig. 13.5). A possible explanation of this phenomenon is the variation in surface tension around the bubble [6]. Evaporation occurs on the hot side of the bubble and condensation on the cold side. The result is a small temperature gradient around the bubble. Since surface tension is an inverse function of temperature, a slightly lower surface tension exists on the outer edge of the bubble. The result is a significant flow of the interface which, in turn, induces a flow of the surrounding liquid. This liquid flow

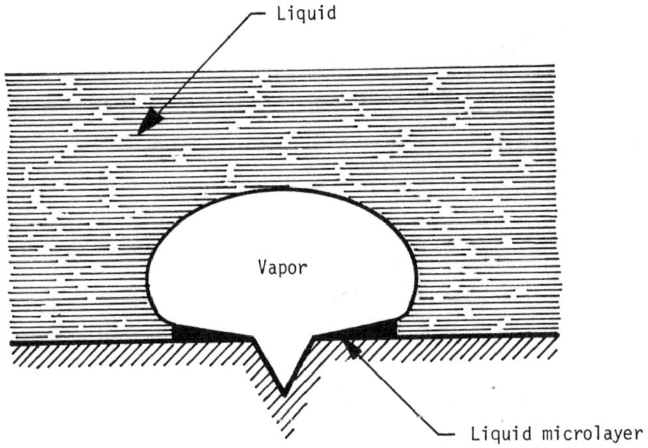

FIGURE 13.4 Model showing formation of a liquid microlayer under a growing bubble.

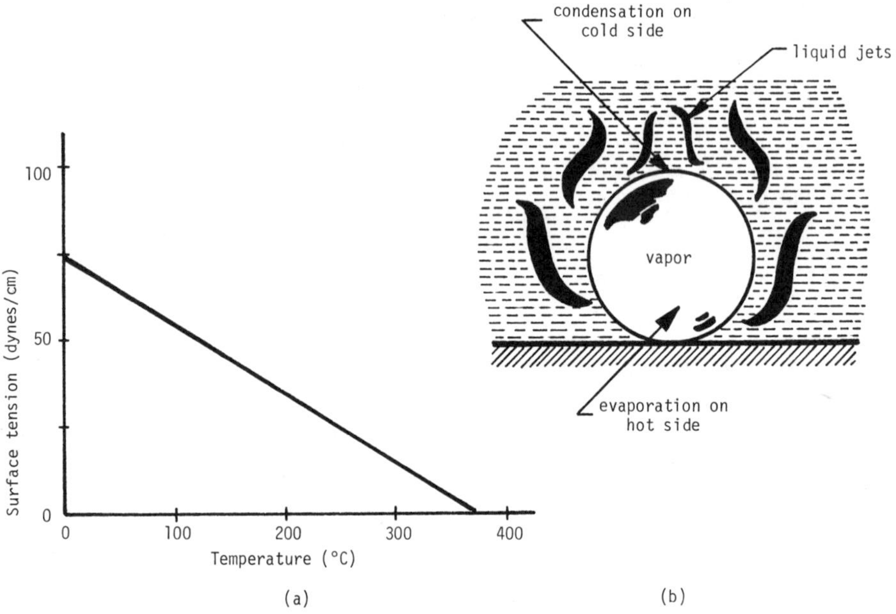

FIGURE 13.5 (a) Surface tension of water as an inverse function of temperature, and (b) model showing jets of liquid resulting from thermocapillarity in subcooled pool boiling.

forms the observed jets, which cool the surface by means of convection currents. The phenomenon is referred to as "thermocapillarity" or "jetting."

13.5 CORRELATIONS FOR BOILING HEAT TRANSFER

The thermal and fluid flow patterns in the various regimes of boiling differ radically, and so it is not possible for a single equation to correlate boiling heat transfer over the complete spectrum.

Important regimes of a boiling system are identified in Fig. 13.2; these regimes are made up of regions which are identified by their intercepts on the curve:

Region	Curve Intercept	Regime
I	1-2	convection
II	2-3	nucleate boiling
III	3-4	nucleate boiling
IV	4-5	nucleate boiling
V	5-6	unstable film boiling
VI	6 →	stable film boiling

Boiling heat transfer has certain similarities to convection heat transfer, and so most boiling correlations are based on the fundamental relationships used for correlating natural and forced convection:

$$\left.\begin{aligned} \mathrm{Nu} &= f(\mathrm{RePr}) \\ &= g(\mathrm{GrPr}) \end{aligned}\right\} \begin{aligned} &\text{forced convection} \\ &\text{natural convection} \end{aligned} \tag{13.3}$$

Region I

In most instances the initial region of pool boiling is one of natural convection. Data can be correlated by using standard natural convection correlations for single-phase systems.

For example, suppose pool boiling occurs from a horizontal square plate with the heated surface facing upward. This condition was studied by Fishenden and Saunders [7] for turbulent and laminar conditions:

turbulent range, $2 \times 10^7 < GrPr < 3 \times 10^{10}$:

$$Nu = 0.14(GrPr)^{1/3} \tag{13.4}$$

laminar range $10^5 < GrPr < 2 \times 10^7$:

$$Nu = 0.54(GrPr)^{1/4} \tag{13.5}$$

In Eqs. (13.4) and (13.5) the characteristic dimension D for the Nusselt and Grashof numbers is the side of the square. Liquid properties are evaluated at the average temperature of the wall and bulk liquid temperature. If the difference between these two temperatures is greater than 5°C, a good practice is to evaluate all liquid properties at the bulk liquid temperature, except viscosity, which should be evaluated at the wall temperature.

Another example of a natural convection correlation that is often used with pool boiling systems is one for natural convection from a horizontal cylinder [8]:

$$Nu = 0.53(GrPr)^{1/4} \tag{13.6}$$

which is valid for the range $10^3 < GrPr < 10^7$. The characteristic dimension is the diameter of the cylinder.

If the initial heat transfer mechanism of the boiling system is one of forced convection, correlations are available in the literature. For example, McAdams [8] recommends the following for liquid flowing normal to a single cylinder for the range $0.1 < Re < 3000$:

$$Nu = 0.34 + 0.56Re^{0.52}\ Pr^{0.3} \tag{13.7}$$

In this correlation the characteristic dimension is the cylinder diameter, and the mass flow velocity is used to evaluate the Reynolds number.

Region III

In the fully developed nucleate boiling region, the heat transfer mechanism is one of violent agitation of the liquid. Therefore, even in pool boiling, the correlation is based on an equation for forced convection. One of the earliest correlations was developed by Rohsenow [6,9]:

$$\frac{c_\ell \, \Delta t_{sat}}{h_{fg}} = C_{sf} \left\{ \frac{q''}{\mu_\ell h_{fg}} \left[\frac{g_c \, \sigma}{g(\rho_\ell - \rho_v)} \right]^{1/2} \right\}^r [\mathrm{Pr}]_\ell^s \tag{13.8}$$

Constants C_{sf}, r, and s must be determined experimentally from pool boiling systems. The value of r for all liquids is taken as 1/3, while s for water = 1.0, and for all other liquids s = 1.71. Constant C_{sf} is a function of a particular heating surface-fluid combination, and has been determined for a number of systems. Table 13.1 lists values of C_{sf} to be used in the Rohsenow correlation for various liquid-surface combinations. Since gravity appears to have no effect on the boiling curve, Rohsenow [6] recommends that $|g| = |g_c|$.

Region II

Since Region II is a transition between convection heat transfer and fully developed nucleate boiling, data for the region can be correlated by a superposition of the two effects. Total heat flux $\Sigma q''$ is expressed as the sum of convective heat flux q_c'' and boiling heat flux q_b'':

$$\Sigma q'' = q_c'' + q_b''$$

Here q_c'' can be obtained from the convection rate equation ($q_c'' = h \, \Delta t$); heat transfer coefficient h can be evaluated from an appropriate correlation such as Eqs. (13.4) to (13.7). Heat flux q_b'' is associated with bubble motion alone without convection and can be calculated from a relationship such as Eq. (13.8).

TABLE 13.1 Values of Constant C_{sf} To Be Used with the Rohsenow Correlation Eq. (13.8) (Exponent r = 1/3 for all liquids, exponent s = 1.0 for water and 1.7 for all other liquids.)

Liquid-surface Combination	C_{sf}
Water-lapped copper	0.0147
Water-chemically etched stainless steel	0.0133
Water-mechanically polished stainless steel	0.0132
Water-platinum	0.013
Water-copper	0.013
Water-polished copper	0.0128
Water-ground polished stainless steel	0.0080
Water-scored copper	0.0068
Water-nickel	0.006
Water-brass	0.006
Water-teflon pitted stainless steel	0.0058
Carbon tetrachloride-copper	0.013
Benzene-chromium	0.010
Carbon tetrachloride-polished copper	0.007
Ethyl alcohol-chromium	0.0027
Isopropyl alcohol-copper	0.0025

Region IV

Correlations for Region IV have not been addressed by most researchers since it is not a usual operating region for commercial equipment. However, the end point of the region has been studied extensively. This point, the peak heat flux, is of great interest because it separates the nucleate and film boiling regimes. Temperatures in film boiling can be orders of magnitude greater than in nucleate boiling. Several correlations have been proposed for peak (or critical) heat flux, Zuber [10]:

$$\frac{q''_{crit}}{\rho_v h_{fg}} = 0.13 \left[\frac{\sigma(\rho_\ell - \rho_v)gg_c}{\rho_v^2}\right]^{1/4} \left[\frac{\rho_\ell}{\rho_\ell + \rho_v}\right]^{1/2} \tag{13.9}$$

Kutateladze [11]:

$$\frac{q''_{crit}}{\rho_v h_{fg}} = 0.16 \left[\frac{\sigma(\rho_\ell - \rho_v)gg_c}{\rho_v^2}\right]^{1/4} \tag{13.10}$$

Rohsenow and Griffith [12]:

$$\frac{q''_{crit}}{\rho_v h_{fg}} = 143 \left(\frac{g}{g_c}\right)^{1/4} \left(\frac{\rho_\ell - \rho_v}{\rho_v}\right)^{0.6} \tag{13.11}$$

The latter equation is not homogeneous but agrees well with most data [6].

Regions V and VI

Region V is one of transition between nucleate and film boiling and is unstable. No generally accepted correlations exist for this region at this time.

In Region VI of a boiling system the heated surface is blanketed by a stable film of vapor. In addition, the temperature of the surface becomes very high so that the heat transfer is a combination of conduction and radiation. The high surface temperatures of film boiling preclude its use in most commercial systems; it does, however, occur in some chemical process equipment and in cryogenic systems.

Bromley proposed an expression for the conductive heat transfer coefficient h_c for film boiling outside a horizontal tube in a pool:

$$h_c = 0.62 \left[\frac{k_v^3 \rho_v(\rho_\ell - \rho_v)h_{fg}}{D\mu_v(T_w - T_{sat})}\right]^{1/4} \tag{13.12}$$

An expression for the radiation heat transfer coefficient h_r can be obtained from the Stefan-Boltzmann equation, modified to allow for effects of geometry and reradiating surfaces:

$$h_r = \frac{\sigma F_s F_\varepsilon (T_w^4 - T_{sat}^4)}{T_w - T_{sat}} \qquad (13.13)$$

where σ is the Stefan-Boltzmann constant, F_s is the shape factor for radiation, and F_ε is the emissivity factor.

13.6 CONDENSATION

Most practical applications of boiling heat transfer depend to some extent on the condensation of vapor to liquid. When a saturated or superheated vapor comes in contact with a surface whose temperature level is lower than the vapor saturation temperature, condensation takes place. This process is the reverse of evaporation, and a quantity of heat is released, which is determined by the latent heat property of the vapor. If the surface is wetted by the liquid, a film is formed and the process is called film condensation. When the surface is not wetted, beads of liquid form on the surface and are held by surface tension forces until they grow large enough to fall from, or flow on, the surface through gravitational attraction. This latter process is called dropwise condensation and is preferred over film condensation because large sections of the surface are directly exposed to the vapor, resulting in higher heat transfer rates. Dropwise condensation may be an order of magnitude more effective than continuous film condensation. In practice, dropwise condensation is not likely to occur because most surfaces are wetted after an extended exposure to a condensing vapor. Attempts have been made to use various coatings and additives to maintain dropwise condensation, but in general they are not successful.

In film condensation the liquid forms a film between the surface and the vapor, which acts as a thermal resistance whose magnitude is proportional to the film thickness. The thickness is influenced by the fluid properties, the surface finish, and the surface geometry. Liquids of high viscosity and surfaces that are rough tend to impede the flow of liquid, and so the film will thicken. Drainage from vertical or inclined surfaces is more rapid than from horizontal surfaces, and so the film is thinner. Therefore, it is evident that, other things being equal, the most effective condenser will be a smooth, vertical surface condensing a low viscosity liquid. An exception to this is the case of a condenser whose vertical height is great; condensate can accumulate on the lower portion of the surface, increasing the thickness of the film.

Condensation Correlations

Nusselt [13] developed mathematical relations to predict the condensation heat-transfer rate as early as 1916. These equations show fair agreement with most experimental data, and where no noncondensable gases exist, the errors are conservative. The reason for this is that Nusselt neglected vapor velocity effects and assumed continuous film condensation. In certain systems the vapor has some velocity, and the process is a combination of dropwise and continuous film condensation. Nusselt suggested the following relationships:

Horizontal tubes

$$h_c = 0.725 \left(\frac{k^3 \rho^2 g h_{fg}}{D\mu \, \Delta t} \right)^{1/4} \tag{13.14}$$

Vertical and inclined surfaces

$$h_c = 0.943 \left(\frac{k^3 \rho^2 g h_{fg} \sin \theta}{L\mu \, \Delta t} \right)^{1/4} \tag{13.15}$$

where h_c = condensation heat transfer coefficient, k = liquid thermal conductivity, ρ = liquid density, g = acceleration of gravity, h_{fg} = latent heat of condensation, D = outside diameter of tube, μ = liquid absolute viscosity, Δt = temperature difference between the saturated vapor and the wall, L = slant height of surface, and θ = angle between surface and horizontal.

The liquid properties in these correlations are to be evaluated at the arithmetic mean of the vapor and wall temperatures. An interesting consequence of Eqs. (13.14) and (13.15) is that the heat transfer coefficient decreases as Δt increases. This is consistent with the increase in thermal resistance that results from the film buildup brought about by the increase in the rate of condensation.

Effects of Noncondensable Gas on Condensation

It is well known that the presence of a noncondensable gas (such as air) in a condensing system will reduce the rate of heat transfer of the process. The mechanism that is responsible for the reduction is shown schematically in Fig. 13.6. The inert noncondensable gas forms a layer on the liquid film through which the vapor must diffuse. There is a drop of partial pressure in the vapor that represents the driving potential for the mass diffusion process. Since the saturation temperature decreases with the pressure, the saturation temperature at the liquid-gas interface will be less than that in the bulk mixture. The effect of the temperature reduction can be appreciated by considering the expression for the heat transfer rate:

$$q = h_m A(t_{sat} - t_w) \tag{13.16}$$

where q is the heat transfer rate to a condenser of area A and mean heat transfer coefficient h_m when the temperature potential is the difference between the saturation temperature t_{sat} and the cold wall temperature t_w. It is evident that the heat transfer rate will be decreased when the saturation temperature is lowered.

Practical Applications

There are innumerable examples of condenser designs to be used with heating, cooling, power generating, processing, and communication equipment. In most cases it will be convenient and practical for the designer to use commercially available heat exchangers; also, the liquid and vapor involved will be selected according to the requirements of the total system, not just for condensation properties. Therefore, the only significant variables under the control of the condenser designer are the temperature and velocity of the secondary fluid coolant. Even these may have practical limitations such as

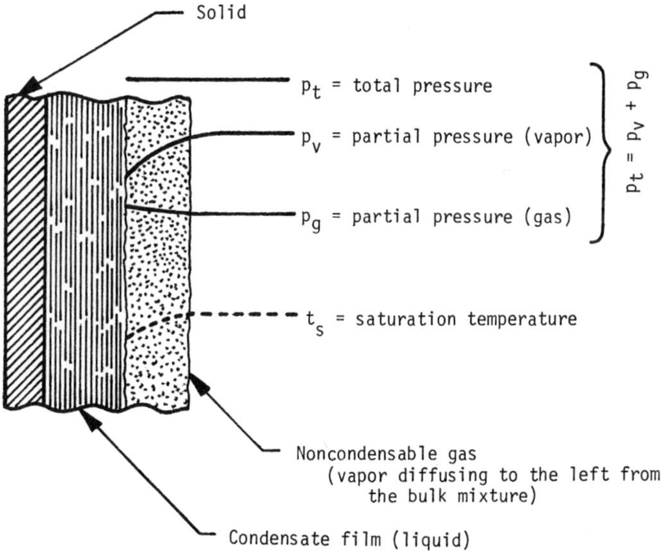

FIGURE 13.6 Model showing reason for decreased heat transfer rate in a condensing system that contains a noncondensable gas.

minimum dew point requirements for the coolant. Coolant velocity may also be limited by pump or fan capacity, duct size, or acoustical noise specifications.

13.7 CONCLUSION

The phenomenon of boiling heat transfer is introduced in this chapter. Basic nomenclature is presented, and the regimes of boiling are identified and shown graphically on a boiling curve. Some of the generally accepted heat transfer mechanisms of nucleate boiling are described and represented schematically. Dimensionless correlations for each boiling regime are presented in forms that can be used for analysis. The concept of condensation is covered briefly, and the effects of noncondensable gases is discussed.

SOLVED PROBLEMS

PROBLEM 1: Water at 1 atm is boiled on a brass surface. Using the Rohsenow correlation, determine the heat transfer coefficient for nucleate boiling if the wall superheat is 11°C.

Solution:

$$\frac{c_\ell \, \Delta t_{sat}}{h_{fg}} = C_{sf} \left\{ \frac{q''}{\mu_\ell h_{fg}} \left[\frac{g_c \, \sigma}{g(\rho_\ell - \rho_v)} \right]^{1/2} \right\}^r (Pr)_\ell^s$$

from Table 13.1:

$$C_{sf} = 0.006$$
$$r = 1/3$$
$$s = 1.0$$

Thermodynamic properties are evaluated at $t_{sat} = 100°C$ for 1 atm (see table A-3, Ref. 14).

$$c_\ell = 4.218 \text{ J/gm K}$$
$$h_{fg} = 2676 - 419 = 2257 \text{ J/gm}$$
$$Pr = 1.75$$
$$\mu_\ell = 283.1 \text{ Ns/m}^2 \text{ or kg/m s} = 283.1 \times 10^{-3} \text{ gm/m s}$$
$$\sigma = 57.78 \times 10^{-3} \text{ N/m}$$
$$\rho_\ell = 1/v_\ell = 1/1.0435 \times 10^{-3} = 958 \text{ kg/m}^3 = 958 \times 10^3 \text{ gm/m}^3$$
$$\rho_v = 1/v_v = 598 \text{ gm/m}^3$$
$$\rho_\ell - \rho_v = 957.4 \times 10^3 \text{ gm/m}^3$$

also,

$$g_c = 10^3 \text{ gm m/N s}^2$$
$$g = 9.8 \text{ m/s}^2$$
$$\Delta t_{sat} = 11°C$$

$$\frac{(4.218)(11)}{(2257)(1.75)^{1.0}} = 0.006 \left\{ \frac{q''}{(283.1 \times 10^{-3})(2257)} \left[\frac{(10^3)(58.78 \times 10^{-3})}{(9.8)(957.4 \times 10^3)} \right]^{1/2} \right\}^{1/3}$$

$$q'' = 1.88 \times 10^6 \text{ W/m}^2$$

$$h = q''/\Delta t_{sat} = 1.88 \times 10^6/11 = 1.71 \times 10^5 \text{ W/m}^2 \text{ °C}$$

PROBLEM 2: A tube, 1/2 in OD and 5 ft long, is used to condense steam at 6 psia; the average temperature of the tube is maintained at 130°F. Determine the condensation heat transfer coefficient if the tube is mounted (a) vertically and (b) horizontally.

Solution: Condensation h.t. coefficients can be found by applying the Nusselt relationships, Eqs. (13.14) and (13.15). From steam tables:

$$T_{sat} \text{ at 6 psia} = 170.03\text{°F}$$

$$h_{fg} = 996.2 \text{ Btu/lbm}$$

Thermodynamic properties are evaluated at the film temperature T_f

$$T_f = \frac{170.03 + 130}{2} = 150\text{°F}$$

$$\Delta T = 170 - 130 = 40\text{°F}$$

$$D = 0.5 \text{ in} = 0.042 \text{ ft}$$

$$L = 5 \text{ ft}$$

$$g = 32.2 \text{ ft/s}^2 = 4.17 \times 10^8 \text{ ft/h}^2$$

From *Principles of Heat Transfer* by Kreith (table A-3):

$$k = 0.384 \text{ Btu/hr ft °F}$$

$$\rho = 61.2 \text{ lbm/ft}^3$$

$$\mu = 0.292 \times 10^{-3} \text{ lbm/ft s} = 1.05 \text{ lbm/ft h}$$

(a) Vertical Orientation

$$h_c = 0.943 \left(\frac{k^3 \rho^2 g h_{fg}}{L\mu \, \Delta T} \right)^{1/4}$$

$$h_c = 0.943 \left[\frac{(0.384)^3(61.2)^2(4.17 \times 10^8)(996.2)}{(5)(1.05)(40)}\right]^{1/4}$$

$$= 759 \text{ Btu/h ft}^2 \text{ °F}$$

(b) Horizontal Orientation

$$h_c = 0.725 \left(\frac{k^3\rho^2 g h_{fg}}{D\mu\ \Delta T}\right)^{1/4}$$

$$h_c = 0.725 \left[\frac{(0.384)^3(61.2)^2(4.17 \times 10^8)(996.2)}{(0.042)(1.05)(40)}\right]^{1/4}$$

$$= 1927 \text{ Btu/h ft}^2 \text{ °F}$$

EXERCISES

1. Define pool boiling and give one practical example.
2. Define flow boiling and give one practical example.
3. What is the essential difference between subcooled and saturated nucleate boiling?
4. If water at saturation temperature and atmospheric pressure is boiled by means of an electrically heated wire maintained at 1412°F, what is the water superheat?
5. Water at 1 atm boils on a brass surface. Using the Rohsenow correlation, determine the heat transfer coefficient for nucleate boiling if the wall superheat is 20°F.
6. What is the difference between film and dropwise condensation?
7. Saturated steam at 216°F condenses on four 1-in diameter, 2-ft long horizontal tubes arranged in a vertical bank. The tube wall is kept at 200°F. Determine the heat transfer rate due to condensation.
8. How does a noncondensable gas, such as air, affect the rate of heat transfer in a condensation process?
9. In most practical cases of condenser design, what are the most important parameters to be considered?

REFERENCES

1. J. Jakob, *Heat Transfer*, Vol. I, Chap. 29, Wiley, New York, 1949.

2. C. Y. Han and P. Griffith, The mechanism of heat transfer in nucleate pool boiling, Part I, Bubble initiation, growth, and departure, *Int. J. Heat Mass Transfer*, Vol. 8, p. 887, 1965.

3. H. K. Forster and R. Greif, Heat transfer to a boiling liquid; mechanism and correlation, Trans. ASME, Ser. C, *J. Heat Transfer*, Vol. 81, pp. 43-53, 1959.

4. F. D. Moore and R. B. Mesler, The measurement of rapid surface temperature fluctuations during nucleate boiling of water, *AIChE J.*, Vol. 7, pp. 620-624, 1961.

5. F. Gunther and F. Kreith, *Photographic Studies of Bubble Formation in Heat Transfer to Subcooled Water*, Jet Propulsion Lab Progress Report 4-120 C.I.T., 1950.

6. W. Rohsenow and J. Hartnett, *Handbook of Heat Transfer*, Chap. 13, McGraw-Hill, New York, 1973.

7. M. Fishender and O. Saunders, *An Introduction to Heat Transfer*, Oxford Univ. Press, Oxford, 1950.

8. W. H. McAdams, *Heat Transmission*, McGraw-Hill, New York, 1954.

9. W. M. Rohsenow, *A Method of Correlating Heat Transfer Data for Surface Boiling of Liquids*, Trans. ASME, July 1952.

10. N. Zuber, *Hydrodynamic Aspects of Boiling Heat Transfer*, USAEC Report AECU-4439, 1959.

11. S. S. Kutateladze, *Heat Transfer in Condensation and Boiling*, USAEC Report AEC-tr 3770, 1952.

12. W. M. Rohsenow and P. Griffith, *Correlation of Maximum Heat Transfer Data for Boiling of Saturated Liquids*, Chem. Engr. Progr. Symp. Series, Vol. 52, No. 18, 47 pp., 1956.

13. W. Nusselt, *Ver. deut. Ing.*, Vol. 60, pp. 541-569, 1916.

14. J. G. Collier, Convective Boiling and Condensation, McGraw-Hill, New York, 1972.

14

RADIATION HEAT TRANSFER

14.1 INTRODUCTION

Heat transfer by radiation differs from heat transfer by conduction and convection in that no medium is needed for the transmittal of radiant thermal energy. Also, radiation heat transfer is proportional to the temperature level of a system--actually the fourth power of absolute temperature--whereas conduction and convection depend upon a temperature difference. Consequently, heat transfer from a high temperature system is due mostly to radiation, while conduction and convection predominate at lower temperatures.

In addition to temperature, radiation heat transfer depends upon the nature of the radiating surface and its position with respect to other sources or sinks.

14.2 PHYSICAL MECHANISM

Radiation can be defined as the continual emission of energy from the surface of a body as a result of the absolute temperature and the properties of the material of the radiating body. Radiation is transmitted in various forms of electromagnetic waves, such as radiowaves, infrared and ultraviolet waves, and x-rays. The distinguishing characteristic among the various waves is the wavelength (or frequency), and the array of waves can be shown on an electromagnetic

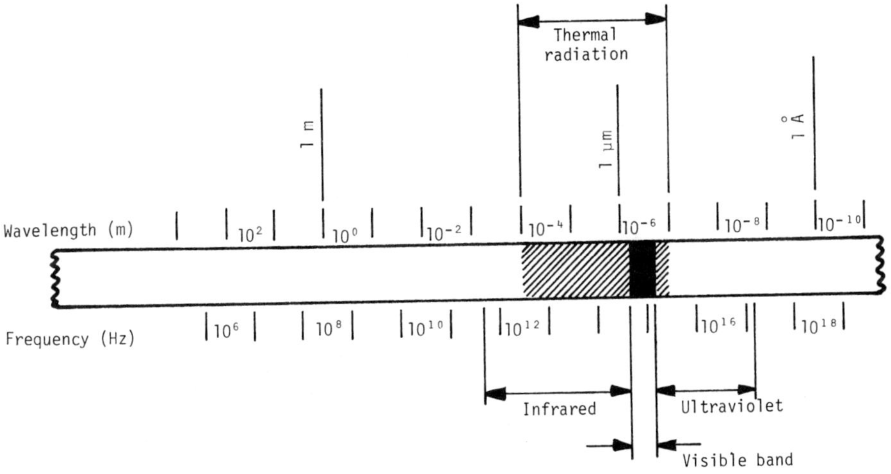

FIGURE 14.1 Electromagnetic wave spectrum. Radiation in the range between 0.1 and 100 μm causes heating that can be detected by humans. Radiation in a narrower band from 0.38 to 0.76 μm stimulates the human optical nerve.

wave spectrum (Fig. 14.1). Radiated waves do not require a medium for propagation, and so they can travel through a vacuum.

Each wave, regardless of type, is propagated in a vacuum at the speed of light ($c = 3.0 \times 10^8$ m/s). This constant velocity is equal to the product of the wavelength λ and the frequency ν:

$$c = \lambda\nu \tag{14.1}$$

The unit for λ is generally taken as the meter (1 m = 3.28 ft), the angstrom ($1\ \text{Å} = 10^{-10}$ m), or the micron ($1\ \mu\text{m} = 10^{-6}$ m). The standard unit for ν is the hertz ($1\ \text{Hz} \equiv 1$ cycle/s).

Radiation phenomenon can be predicted approximately by numerical analysis, but its true nature has not been fully established. Analysis is based on two contradictory theories: wave theory and quantum theory.

The wave theory provides a qualitative explanation of radiant energy transfer. When a body emits radiant energy, a portion of its internal energy is converted into electromagnetic waves. These waves, which are forms of energy, travel through space until they strike another body where some portion of the energy is reconverted into internal energy. The loss of internal energy by the emitting body is evidenced by a reduction in temperature and, conversely, the temperature of the receiving body rises.

The quantum theory provides another simplified view of radiation propagation. A quantum can be considered to have energy, mass, and momentum in the same manner as the molecules of a gas possess these characteristics. Consequently, radiation can be thought of as a "photon gas" that can flow from place to place. This analogy permits the formulation of several equations that can be used for the analysis of radiation phenomena.

The first of these equations was developed by Max Planck in 1900 as an empirical relation that satisfactorily represented the energy distribution observed in the spectrum of an ideal radiating body. It states that the propagation of thermal radiation takes place in the form of discrete quanta, each quantum having an amount of energy E expressed as

$$E = h\nu \tag{14.2}$$

The proportionality constant has a value $h = 6.62 \times 10^{-34}$ J s. With this relation and the relativistic relation between mass and energy, expressions for mass and momentum of particles in the "photon gas" can be derived:

$$E = mc^2 = h\nu$$

$$\therefore \quad m = \frac{h\nu}{c^2} \tag{14.3}$$

and

$$\text{Momentum} = cm = \frac{ch\nu}{c^2} = \frac{h\nu}{c} \tag{14.4}$$

14.3 ABSORPTION, REFLECTION, AND TRANSMISSION

When radiant energy strikes a surface, a portion of it is absorbed, a portion is reflected, and a portion is transmitted as shown in Fig. 14.2. It is evident that the following relation holds for the three conditions:

$$\alpha + \rho + \tau = 1.0 \tag{14.5}$$

where

α = absorptivity (the fraction absorbed)

ρ = reflectivity (the fraction reflected)

τ = transmissivity (the fraction transmitted)

Most solids are opaque, that is, they do not transmit radiation, instead they absorb it within a thin surface layer. This layer is on the order of 0.1 cm. For solids, then, the transmissivity is zero and the following relation holds:

$$\alpha + \rho = 1.0 \tag{14.6}$$

When radiation strikes a surface two kinds of reflection can be noted. The first type occurs on highly polished, smooth surfaces. The angle at which the ray strikes the surface (*angle of incidence*) is equal to the angle at which it leaves the surface (*angle of reflection*). This type of reflection is called *specular* or *regular reflection* and is encountered mostly in optical systems. Most practical

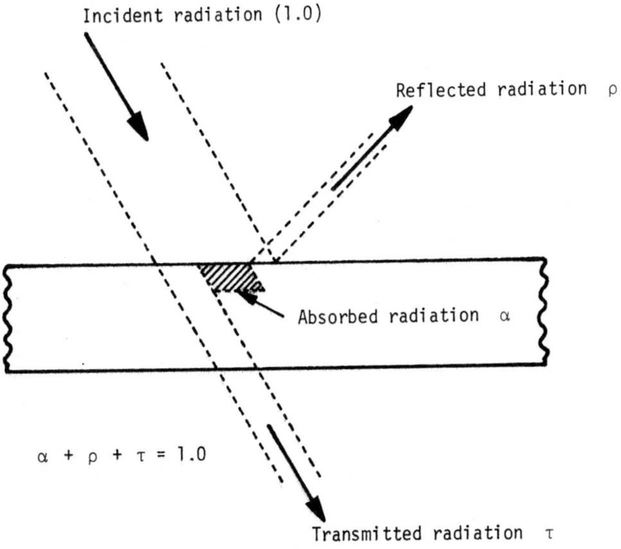

FIGURE 14.2 Reflected, absorbed, and transmitted radiant energy make up the incident radiation.

materials, however, can be considered rough. This means that the surface asperities are large compared with the wavelength of the wave. When this condition obtains, radiation is reflected indiscriminately in all directions and is called *diffuse reflection.* Since radiation is influenced by the wavelength of the incident wave, it is possible that a surface might be a good reflector of visible light but a poor reflector of infrared waves, which have longer wavelengths.

14.4 BLACKBODY CONCEPT

Radiant energy impinges on the surface of a body from the inside as well as the outside. If we consider a solid where the transmissivity is zero and Eq. (14.6) holds, then radiant energy will be reflected and absorbed from both sides of the surface. When $\alpha > \rho$, a larger fraction of energy will be absorbed and a smaller fraction reflected from an outside incident wave; the same distribution will be associated with rays hitting the surface from inside the body. A surface of this type will be a poor reflector and a good absorber of energy; it will also be a good emitter. Conversely, if $\alpha < \rho$, the surface will be a good reflector, a poor absorber, and a poor emitter. These relationships are shown schematically in Fig. 14.3a and b.

If it were possible to have a surface where ρ and τ were zero, then all the incident energy striking the surface from outside would be absorbed and, correspondingly, all the incident energy from outside would be emitted. Such a surface would be an ideal absorber and emitter and it is called a *blackbody* surface. A schematic representation of such a surface is shown in Fig. 14.3c.

The name blackbody is used because when such a body is viewed by reflected light only, it appears black. The terminology is not precise because, although a blackbody does not reflect light energy, it does radiate; consequently, if the temperature is high enough a radiating blackbody will glow red or even white.

A blackbody is an ideal concept and does not occur in nature. It is approximated very closely, however, by a *hohlraum* (Fig. 14.4).

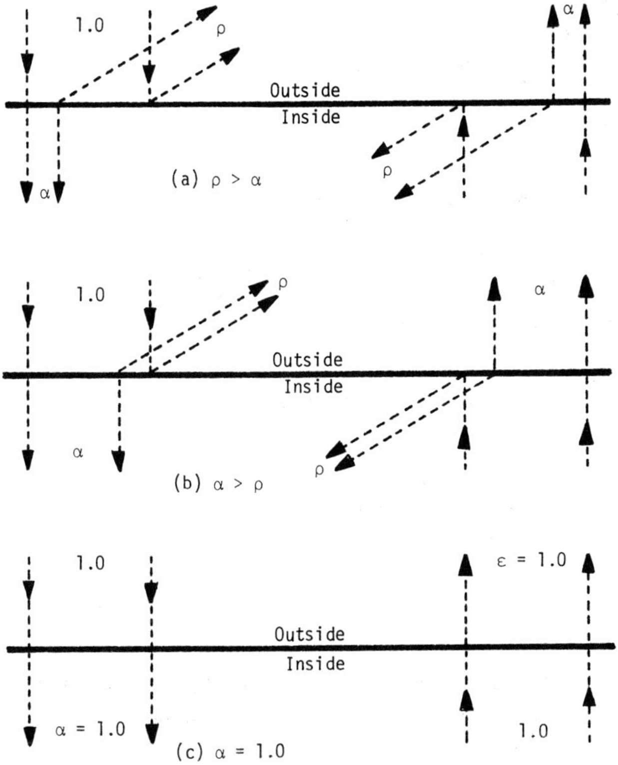

FIGURE 14.3 Distribution of incident radiation: (a) surface is a good reflector, good absorber, and poor emitter, (b) surface is a poor reflector, good absorber, and good emitter, and (c) surface represents a blackbody.

This is an enclosure with a small opening. Radiation streams through the opening and is reflected repeatedly and absorbed by the walls. In this manner practically all the incoming energy is absorbed. Consequently, the opening in a hohlraum can be considered to be a blackbody. If the space is heated, the hohlraum emits blackbody radiation; this is the principle upon which the infrared microscope operates in measuring temperature.

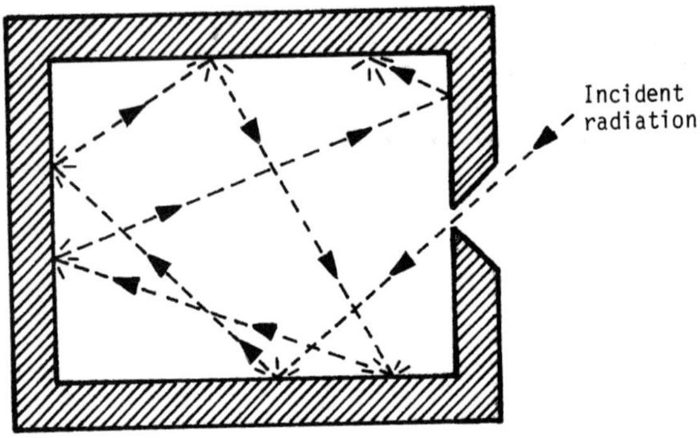

FIGURE 14.4 Hohlraum: small opening in the walls of an enclosure approximates a blackbody. Incident radiation is repeatedly reflected and absorbed by inner walls so that very little energy escapes through the opening. If the enclosure is heated to a temperature above its environment, the opening approximates a blackbody emitter.

14.5 KIRCHHOFF'S LAW

Kirchhoff's law states that, at thermal equilibrium, the ratio of the emissive power of a surface to its absorptivity is constant for all bodies.

Emissive power W is the radiant energy E per unit time and per unit area, and the basic unit for W is 1 J/sm^2 or 1 W/m^2.

Proof of Kirchhoff's law can be shown by considering an enclosure (Fig. 14.5) which is a perfectly insulated vacuum and whose walls are at a uniform temperature. Two small bodies of surface area A_1 and A_2 are placed within the enclosure; radiation is transferred until thermal equilibrium is reached. At that point each body emits just as much energy as it receives. If the energy rate per unit surface area (*irradiation*) from the walls to the bodies is G, and if the absorptivities of the two bodies are α_1 and α_2, then the emissive powers W_1 and W_2 can be used to write energy balances:

$$A_1 G \alpha_1 = A_1 W_1$$

$$A_2 G \alpha_2 = A_2 W_2$$

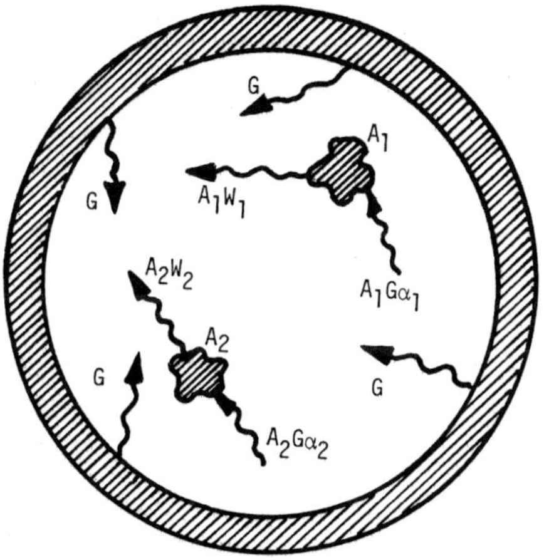

FIGURE 14.5 Model to illustrate Kirchhoff's law: large, perfectly insulated, evacuated enclosure with energy radiating from the walls at a rate G. Energies absorbed and emitted from two bodies of area A_1 and A_2 are equal at thermal equilibrium.

or

$$\frac{W_1}{\alpha_1} = \frac{W_2}{\alpha_2} = \frac{W}{\alpha} \tag{14.7}$$

Equation (14.7) is the mathematical statement of Kirchhoff's law.

Since a blackbody has $\alpha = 1.0$, it is evident that a blackbody has the greatest possible emissive power. Also, the ratio of the emissive power W of a given body to the emissive power W_b of a blackbody, when both bodies are at the same temperature, is equal to the absorptivity of the given body:

$$\frac{W}{W_b} = \alpha \tag{14.8}$$

This ratio defines the emissivity ε, which is a measure of the effectiveness of a surface as a radiator relative to that of a blackbody:

$$\varepsilon \equiv \frac{W}{W_b} \tag{14.9}$$

Therefore, a corollary to Kirchhoff's law is the Kirchhoff identity:

$$\varepsilon = \alpha \tag{14.10}$$

For blackbodies, ε and α have a value of 1.0, while for all real bodies the value lies in the interval between 0 and 1.0.

Emissivity depends on temperature and on the surface condition of the radiating body. For example, some metals when cleaned and polished have very low emissivities, whereas the same metal when oxidized will exhibit large values of ε. Table 14.1 lists some average values of ε for several materials at ordinary temperatures (≈40°C).

In most practical cases, a body emits most of its radiation at a temperature which is different from that at which it received the radiation. Consequently, the average values of ε and α are not the same. The average absorptivity α should correspond to the temperature at which an incoming radiation was emitted, while the average ε should correspond to the temperature of the emitting body being analyzed.

14.6 PLANCK'S LAW

Planck's hypothesis, given by Eq. (13.16), marked the beginning of the quantum theory. In words it states that the atomic energy of a

TABLE 4.1 Emissivities of Various Surfaces at Approximately 40°C

MATERIAL	ε
Polished Silver	0.01
Polished Aluminum	0.04
Polished Copper	0.04
Oxidized Copper	0.87
Polished Stainless Steel	0.15
Red Brick	0.93
White Paper	0.95
Lampblack Paint	0.96
Ice	0.97

system can have only discrete sets of values. The energy can increase or decrease, but only in a discontinuous, stepwise fashion.

Planck's law evolved from his original hypothesis and is a relationship which shows how the emissive power of a blackbody is distributed among different frequencies and wavelengths. The law is given here without proof.

Let $W_{b\lambda}$ be the monochromatic emissive power per unit wavelength for a blackbody; that is, the radiant energy per unit time and per unit area emitted from a blackbody at some particular wavelength. Planck's law states that $W_{b\lambda}$ has a value of

$$W_{b\lambda} = \frac{2\pi hc^2\lambda^{-5}}{e^{ch/k\lambda T} - 1} \tag{14.11}$$

where

$W_{b\lambda}$ has units of W/m^3

$h = 6.62 \times 10^{-34}$ Js (Planck's constant)

$c = 2.998 \times 10^8$ m/s (speed of light)

λ = wavelength in meters

$e = 2.71$ (napierian base)

$k = 1.38 \times 10^{-23}$ J/K (Boltzmann constant)

T = absolute temperature of emitting surface (K)

Since Eq. (14.11) contains a number of constants, it is usually stated in the form

$$W_{b\lambda} = \frac{C_1\lambda^{-5}}{e^{C_2/\lambda T} - 1} \tag{14.12}$$

where

$C_1 = 3.74 \times 10^{-16}$ Jm^2/s

$C_2 = 1.44 \times 10^{-2}$ mK

A plot of Eq. (14.11) is shown in Fig. 14.6 with $W_{b\lambda}$ and λ as variables and absolute temperature T a parameter. Note that the maximum value of $W_{b\lambda}$ is displaced toward shorter wavelengths as the

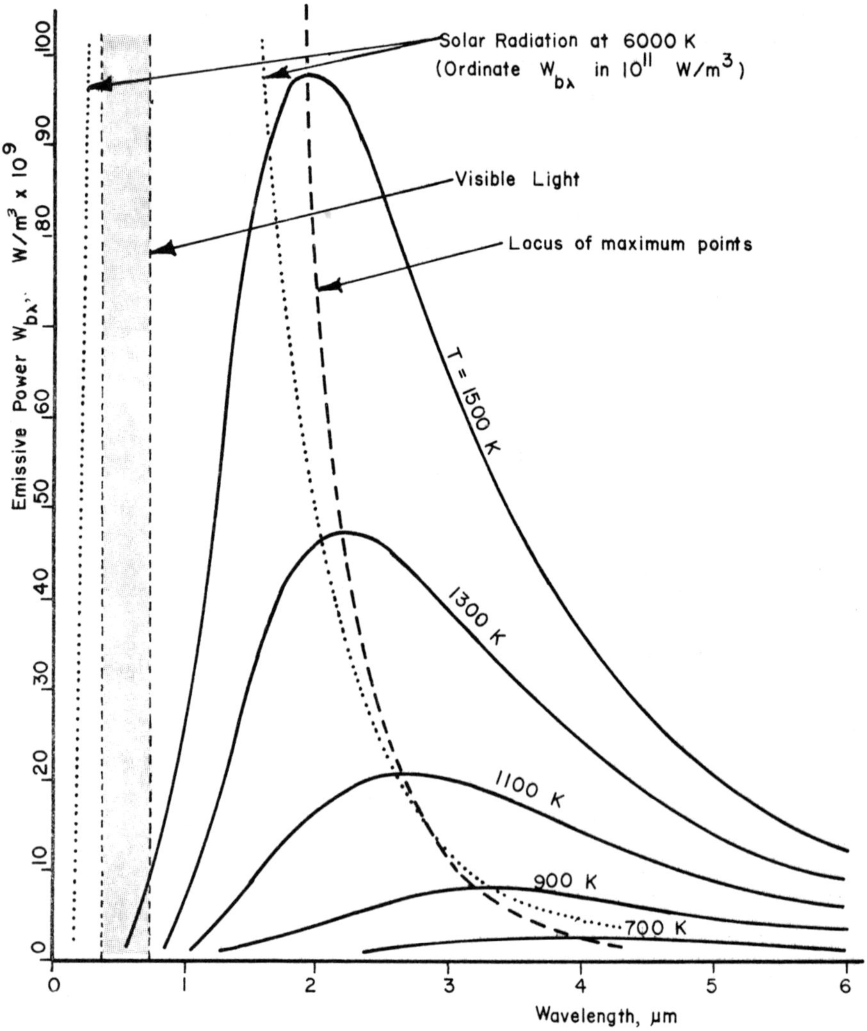

FIGURE 14.6 Monochromatic emissive power distribution for a blackbody at various absolute temperatures. Spectroradiometric curves are plotted from Planck's law [Eq. (14.11)].

temperature increases. This inverse relationship between wavelength and temperature was discovered experimentally by Wien before Planck's law was developed. It is known as Wien's displacement law and has the form

$$\lambda_{max} T = 2893 \tag{14.13}$$

where λ_{max} is the wavelength in microns at which the value of $W_{b\lambda}$ is a maximum when the emitting surface is at an absolute temperature in degrees K.

The total radiant energy W emitted by a plane surface per unit area and unit time can be found by integrating Eq. (14.12) with respect to wavelength λ:

$$W = \int_0^\infty W_{b\lambda}\, d\lambda \tag{14.14}$$

and is represented graphically as the area under the curves of Fig. 14.6. This allows one to determine the range of wavelengths (or frequencies) over which most of the radiant energy is emitted from a surface of a particular temperature. For example, Fig. 14.6 shows that most of the sun's radiation is emitted between wavelengths of 0.1 and 3.0 μm. For surfaces at lower temperatures, it can be seen that the bulk of emissions occurs at longer wavelengths. This phenomenon results in the *greenhouse effect*. Glass is transparent to the short wavelengths of the sun's radiation and, consequently, heat energy enters the greenhouse. Glass, however, is opaque to the longer wavelengths emitted from the lower temperature surfaces inside the greenhouse so that the heat energy cannot escape. As a result, the temperature inside the greenhouse attains a higher value than the air outside.

Figure 14.6 illustrates another phenomenon of radiant energy. As the temperature of a body increases, a larger amount of radiation becomes visible to the human eye. This is the principle used by the blacksmith to estimate the temperature of metal for a tempering process. At approximately 1000 K, radiant energy is transmitted with a dull red color. As the temperature increases, the color proceeds through shades of cherry red and straw yellow and, finally, at about 3000 K the color is very nearly white.

An interesting observation related to the phenomena under consideration has to do with the evolution of the human eye. During its development, the eye has become sensitive to wavelengths of light over which the sun radiates most of its energy.

Much of the heat transfer literature is written in British units, and so the constants in Planck's law [Eq. (14.11)] often appear as

$$C_1 = 1.19 \times 10^8 \text{ Btu } \mu\text{m}^4/\text{hr ft}^2$$

$$C_2 = 2.59 \times 10^4 \text{ } \mu\text{m}°\text{R}$$

When these constants are used, $W_{b\lambda}$ has units of Btu/hr ft^2 μm.

Also, the Wien displacement law when expressed in British units has the form

$$\lambda_{max}T = 5216 \tag{14.15}$$

where λ is in microns and T is the absolute temperature in degrees R.

14.7 STEFAN-BOLTZMANN LAW

The Stefan-Boltzmann law expresses the total emissive power from a blackbody as a function of the fourth power of absolute temperature. The relation, as stated, includes all wavelengths and frequencies:

$$W_b = \sigma T^4 \tag{14.16}$$

Several approximate values for the proportionality constant are given below in a variety of units:

$$\sigma = 1.36 \times 10^{-12} \text{ cal/s cm}^2 \text{ K}^4$$

$$= 5.7 \times 10^{-5} \text{ erg/s cm}^2 \text{ K}^4$$

$$= 5.7 \times 10^{-8} \text{ W/m}^2 \text{ K}^4$$

$$= 0.17 \times 10^{-8} \text{ Btu/hr ft}^2 \text{°R}^4$$

The law expressed by Eq. (14.16) was developed experimentally by Stefan and analytically by Boltzmann prior to the development of Planck's law. It can, however, be derived from Planck's law.

14.8 GRAY BODY CONCEPT

By definition the ratio of the monochromatic emissive power of a *gray body* to the monochromatic emissive power of a blackbody at the same wavelength is constant over the entire spectrum. This means that the monochromatic emissivity ε_g for a gray body is defined as

$$\varepsilon_g \equiv \frac{W_\lambda}{W_{b\lambda}} \tag{14.17}$$

Also, the Stefan-Boltzmann law [Eq. (14.16)] is modified for a gray body:

$$W_g = \varepsilon_g \sigma T^4 \tag{14.18}$$

The shape of the spectroradiometric curves for a gray body are similar to those for a blackbody, as shown in Fig. 14.6, except that the amplitudes are everywhere reduced by the numerical value of the emissivity.

The concept of the gray body is less idealistic than that of a blackbody, but it is still not completely realistic. Many surfaces deviate from gray body specifications, namely, the emissivity is independent of wavelength. In practical heat transfer work, however, the assumption is usually made that a body is either a blackbody or a gray body.

14.9 RADIATION HEAT TRANSFER BETWEEN SURFACES

Equations developed up to this point refer to emission or absorption of individual surfaces and not to any net energy exchange between two or more surfaces. Of course, such an exchange is of interest to the heat transfer engineer working on analysis of a heat transfer system. Furthermore, heat transfer engineers do not generally use emissive power W when considering heat flow. Instead, they refer to *heat transfer rate* q. The relationship between W and q is given by

$$q = WA \tag{14.19}$$

where q has dimensions of heat energy per unit time and A is the area of the radiating body.

Suppose, for example, that a body of area A_1 and temperature T_1 is transferring blackbody radiant energy to the walls of an enclosure that completely surrounds it; the walls of the enclosure are at temperature T_2. It should be evident that application of the Stefan-Boltzmann equation and Eq. (14.19) will lead to an expression that gives the net loss of energy of body A_1:

$$q = A_1\sigma(T_1^4 - T_2^4) \tag{14.20}$$

Blackbody Shape Factor

The derivation of Eq. (14.20) is based on the assumption that all of the radiant energy emitted from A_1 reaches the body whose temperature is T_2. When A_1 is completely enclosed by A_2, this assumption is obviously valid. There are, however, various geometric arrangements of A_1 and A_2 where radiant energy from one body will not reach the other body. To account for this possibility, a *shape factor* F is assigned as a measure of how well bodies can "see each other":

F_{1-2} ≡ portion of energy leaving surface 1 which reaches surface 2

F_{2-1} ≡ portion of energy leaving surface 2 which reaches surface 1

Other names are assigned to F in the heat transfer literature, such as *configuration factor*, *view factor*, and *angle factor*.

Shape factor F can be calculated for various configurations of area, but a complex double integration is necessary. Fortunately, F has a value of one for several cases of practical interest. For more complicated configurations, tables and diagrams are available to aid in selecting appropriate values of F. Table 14.2 can be used to determine F for a number of simple cases; other configurations can be found in Refs. 1 and 2.

TABLE 14.2 Geometric Shape Factors and Emissivity Factors to be Used in Eq. (14.23)

SURFACES	AREA	F	F_ε
1. Infinite parallel planes	A_1 or A_2	1.0	$\dfrac{1}{\dfrac{1}{\varepsilon_1} + \dfrac{1}{\varepsilon_2} - 1}$
2. Convex body A_1 completely enclosed in body A_2; body A_1 small with respect to A_2	A_1	1.0	ε_1
3. Convex body A_1 completely enclosed in body A_2; body A_1 large with respect to A_2	A_1	1.0	$\dfrac{1}{\dfrac{1}{\varepsilon_1} + \dfrac{1}{\varepsilon_2} - 1}$
4. Concentric spheres, A_1 inside A_2	A_1	1.0	$\dfrac{1}{\dfrac{1}{\varepsilon_1} + \dfrac{A_1}{A_2}\left(\dfrac{1}{\varepsilon_2} - 1\right)}$
5. Infinite concentric cylinders, A_1 inside A_2	A_1	1.0	$\dfrac{1}{\dfrac{1}{\varepsilon_1} + \dfrac{A_1}{A_2}\left(\dfrac{1}{\varepsilon_2} - 1\right)}$
6. Finite parallel rectangles	A_1		$\varepsilon_1\varepsilon_2$
7. Perpendicular rectangles	A_1		$\varepsilon_1\varepsilon_2$
8. Parallel Disks	A_1		$\varepsilon_1\varepsilon_2$

Reciprocity Theorem

Consider two blackbody surfaces where the warmer body A_1 is transferring radiant heat energy to a cooler body A_2. The expression for the net rate of heat transfer from body 1 to body 2 is

$$q_{1-2} = A_1F_{1-2}\sigma(T_1^4 - T_2^4)$$

A similar expression can be written which is based on area A_2:

$$q_{2-1} = A_2F_{2-1}\sigma(T_2^4 - T_1^4)$$

obviously,

$$q_{1-2} = -q_{2-1}$$

so that,

$$A_1F_{1-2} = A_2F_{2-1} \tag{14.21}$$

This is designated as the *reciprocity theorem* and it shows that the radiant heat dissipated by a source is equal to the radiant heat gained by a sink. The reciprocity is useful when the shape factor for heat transfer in one direction is much simpler than the shape factor in the reverse direction. For example, consider a body of area A_2 enclosed by area A_1, where $T_1 > T_2$. The net heat transfer rate can be found as

$$q_{1-2} = A_1F_{1-2}\sigma(T_1^{\,4} - T_2^{\,4})$$

but in this expression the value of F_{1-2} is not readily known, while F_{2-1} is given as 1.0 in Table 14.2. Therefore, a more convenient way to solve for q is

$$q_{2-1} = A_2F_{2-1}\sigma(T_2^{\,4} - T_1^{\,4})$$

Nonblackbody Shape Factor

Radiant energy exchange between blackbody surfaces is relatively easy to analyze because all incident energy is absorbed on contact and, consequently, only geometric relations need be considered. When nonblackbodies are involved, however, complexities arise because the incident energy is reflected and rereflected. Geometric relations are still important but, in addition, emissivities of the radiating surfaces must also be considered. If a number of simplifying assumptions are made, it is possible to calculate a shape factor for nonblackbody radiation. These assumptions include uniform surface temperatures, diffuse reflections, constant emissivities and reflectivities over all surfaces, and zero transmissivity. The shape factor F_ε calculated with these assumptions is generally called the *emissivity factor*; several values of F_ε for simple cases are shown in Table 14.2.

In some heat transfer literature reference is made to a *gray body shape factor* F_g, which is defined as the product of the geometric shape factor F and the emissivity factor F_ε:

$$F_g \equiv FF_\varepsilon \tag{14.22}$$

Complete Radiation Equation

It is evident from the preceding sections that the radiation heat transfer rate between two surfaces can be found from the Stefan-Boltzmann equation, but the equation must be modified to allow for effects of geometry and for reradiating surfaces. The complete equation for the radiation heat transfer rate from body A_1 to body A_2 has the general form

$$q_{1-2} = A_1F_{1-2}F_\varepsilon\sigma(T_1^4 - T_2^4) \tag{14.23}$$

In many heat transfer analyses, blackbody radiation is assumed so that the calculation will be simpler. Most heat transfer analysts agree that reflection can be neglected without introducing large errors when the absorptivity exceeds a value of 0.9. There are, however, many practical cases where low values of absorptivity are an integral part of the design, so that nonblackbody radiation must be considered in the analysis. For example, the inner walls of a thermos bottle are silvered to reduce heat flow by radiation. In another case, thermocouples used to measure high temperature are surrounded by radiation shields to prevent specious readings.

Radiation Combined with Convection

It was pointed out previously that, for high temperatures, radiation is often the predominating mode of heat transfer, while at lower temperatures it can often be neglected. There are, however, analyses where both convection and radiation must be considered together. For such cases it is convenient to define a radiation heat transfer coefficient h_r:

$$q = h_rA_1(T_1 - T_2) \tag{14.24}$$

where q can be found from Eq. (14.24) as

$$q = h_r A_1 (T_1 - T_2) = A_1 F_{1-2} F_\varepsilon \sigma (T_1^4 - T_2^4)$$

from which

$$h_r = \frac{q}{A_1 (T_1 - T_2)} = F_{1-2} F_\varepsilon \left[\frac{\sigma (T_1^4 - T_2^4)}{T_1 - T_2} \right] \qquad (14.25)$$

The expression in brackets is called the temperature factor F_T and can be found in Ref. 1. A simplified expression for the radiation heat transfer coefficient can then be written as

$$h_r = F_{1-2} F_\varepsilon F_T = F_g F_T \qquad (14.26)$$

The value of h_r is additive with the local value of the convective heat transfer coefficient, and so a total heat transfer coefficient h_Σ can be defined:

$$h_\Sigma \equiv h_c + h_r \qquad (14.27)$$

Coefficient h_Σ can be used to calculate a thermal resistance which, in turn, can be used to establish an electrical analog.

14.10 SOLAR ENERGY SYSTEMS

The sun is the most important body in the solar system since it supplies the energy necessary for the support of life in all other bodies of the system. On earth, awareness of solar energy is experienced most generally through its light producing properties, the process of photosynthesis, and through the use of fossil fuels. In recent years much thought has been given to direct utilization of solar energy for heating systems and power production. Equipment needed to harness solar energy directly can be classified in three types: flat plate collectors, solar furnaces, and solar batteries. The first two types work on the principle of heating a working substance by means of solar energy, while the last one uses photoelectric or thermoelectric principles. Although each of these types differ significantly in construction and application, they all operate from energy supplied through solar radiation.

Solar Radiation

The sun can be considered to be a sphere of 1 390 444 km (diam.). The sun radiates energy over a range of wavelengths from infrared to ultraviolet. Its surface temperature is about 6000 K, and so the maximum intensity of radiation occurs at a wavelength of 5000 Å, which is in the visible green portion of the spectrum. Earth's atmosphere absorbs much of the radiation at each end of the solar spectrum, and so the solar radiation that reaches earth is confined to the visible and near infrared regions, with just a little ultraviolet. This is a fortuitous occurrence because human life could not survive the impact of full range solar radiation.

Considering the average surface temperature of the sun to be 6000 K, the total radiant energy emitted is in the order of 7×10^7 W/m^2. Such tremendous energy generation can only result from nuclear reactions. The current theory is that the sun produces energy through a series of fusion reactions whereby helium nuclei are formed from photons.

Solar Constant

The *solar constant* is defined as the quantity of energy that falls on a unit area in unit time when the area is placed normal to the sun's rays and located on the outer fringes of the earth's atmosphere. The mean value of the solar constant is 1.34×10^3 W/m^2. The rate at which energy reaches the earth's surface, however, is substantially lower than this figure because the radiant energy must pass through a 145 km layer of atmosphere. Particles of water vapor, carbon dioxide, and dust absorb and scatter much of the energy.

The total amount of radiation received by any particular area on earth has a complex dependence on a number of variables, such as latitude, longitude, time of day and year, climatic conditions, and tilt of the surface. Exact solutions cannot be obtained since the constantly changing relative position of earth and sun results in non-steady state conditions; in addition, the temperatures of surroundings are also variable. Corresponding times and locations for the sun's coordinates can be obtained from ephemerides and almanacs. (An *ephemeris* is a table showing the position of a heavenly body on a number of dates in regular sequence.)

Solar Energy Collectors (Flat Plate Type)

A flat plate solar energy collector consists of a metal plate above which air is trapped by one or more sheets of transparent material. Conduits are mounted in close thermal contact with the metal plate to carry the medium to be heated. The best material for the metal plate is copper whose surface has been blackened with paint or by deposition of a black metallic film. If weight and cost are particularly important, the plate can be made of aluminum. The most widely used transparent material is glass and the medium to be heated is generally water. Figure 14.7 is a schematic of typical construction for a flat plate solar energy collector.

The trapped air in the collector acts as an insulator, thereby reducing heat loss by convection. The glass plate is transparent to the short wavelengths of solar radiation but opaque to the long wavelength radiation emitted from the collector surface. This condition can be recognized as the "greenhouse effect" previously described.

Much more engineering design must be applied to a flat plate collector than is shown in Fig. 14.7. For example, the metal plate is usually mounted on some type of building board and thermal insulation is applied to the backside of the board and to the water headers. Rectangular conduits fitted with fastening straps are used so that optimum thermal contact can be achieved. Piping design is not greatly different from conventional heating systems. Several different types of piping layouts may be used, including grid, split grid, sinuous, or grid-sinuous networks. In areas where freezing temperatures can be encountered, an antifreeze such as ethylene glycol is used. All antifreeze solutions have lower specific heats than water, and so corrections must be made to allow for the difference in total heat transfer rate.

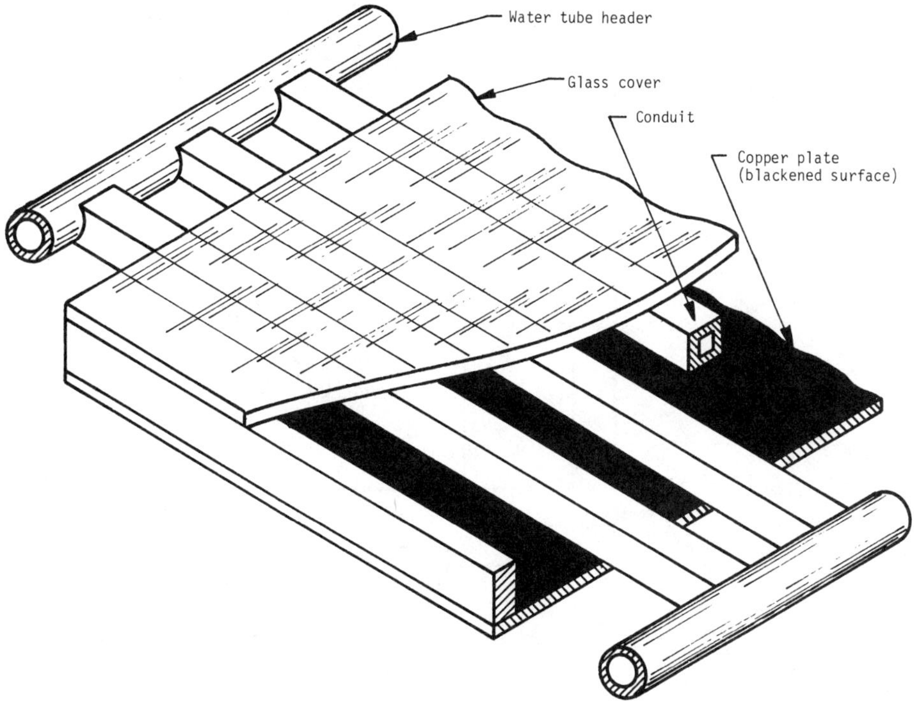

FIGURE 14.7 Schematic of a flat plate solar energy collector.

Solar Heating and Hot Water Systems

The potential value of solar energy for heating homes and providing supplies of hot water has been well known for a long time. Until recently, however, it has not been economically feasible to use solar energy in the construction industry. Recent advances in commercial flat plate collectors, coupled with shortages and soaring costs of fuel, have changed the picture. Collectors are now available which are efficient, easy to install and maintain, and which can be purchased at reasonable costs. In addition, they can be used on new or existing buildings. A well-designed flat plate solar energy collector can heat a fluid more than 60 C above the surrounding air temperature. This makes it practical to use them for heating and air-conditioning homes, heating swimming pools, and providing hot water supplies.

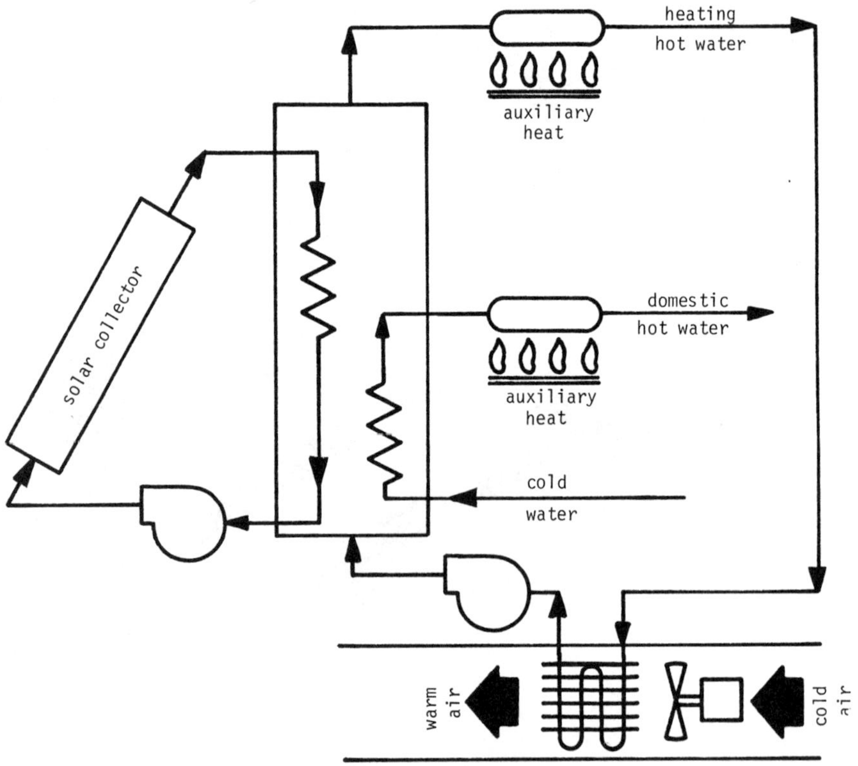

FIGURE 14.8 Solar heating and hot water system.

A schematic drawing of a typical system for heating and providing domestic hot water is shown in Fig. 14.8. Auxiliary heaters are customarily used to allow for variations in incident solar radiation. If the storage tank can be installed high enough, a gravity circulation system can be utilized. A circulating pump is recommended, however, to provide reliable service. Care must be taken to guard against leaks and to ensure that the domestic water is not contaminated.

The most effective performance from a solar collector can be achieved if it is oriented toward the south and at such a slope that it is perpendicular to the sun's rays. Small variations from a due south direction are not significant but the slope of the collector relative to the sun's rays is important.

Local climatic conditions also affect the amount of solar energy that is available. The most important factor is the ratio of sunny days to cloudy days. The U.S. Weather Bureau publishes solar charts and insolation maps which can be very helpful to anyone contemplating the use of solar energy as a source of heating or cooling. Figure 14.9 shows the suitability for the use of solar collection in various parts of the United States.

Other parameters that affect overall performance are surface area of the collector, spacing of the water tubes, and the number of transparent covers used. One, two, or even three covers can be used to reduce the loss of heat. The total number of variables involved in the use of a solar collector makes direct calculation an extremely complex operation. Fortunately, various charts and curves are published in architectural and building construction literature that make analysis relatively simple. Manufacturers of solar collectors also provide similar data for their products.

FIGURE 14.9 Relative suitability for solar energy collection in the continental United States.

Solar Furnace

When high temperatures are required from a solar heating system, a flat plate collector will not suffice, and so a solar furnace must be used. A solar furnace is a paraboloidal concentrator which focuses the sun's rays on a small area located at the focal point. It is necessary that a paraboloidal surface follow the motion of the sun so that it always receives direct rays. This can be done by either moving the reflector itself, or by means of a heliostat.

An important advantage of a solar furnace is that the area from which heat is lost is much smaller than the area which receives the initial radiation. In this way, the heat losses are proportionally reduced and the resulting equilibrium temperature is very high. A well-designed solar furnace can achieve temperature levels up to 4000 K.

The most important applications of the solar furnace occur in high temperature research. Heating by solar radiation instead of combustion or electricity permits the achievement of high temperatures without the contaminating effects of flames, combustion products, and electrodes. Other names for the solar furnace which often appear in the literature are solar motor and solar engine.

Solar Battery

A solar battery is made up of solar cells which, in turn, are photovoltaic diodes that are capable of converting solar energy to electrical energy. The phenomenon involved is called the *photoelectric effect*, which was suggested by Einstein in 1905 based on prior work done by Planck. Einstein postulated that a beam of light consisted of small bundles of energy (quanta or photons) whose energy level could be given by the following relation:

$$E = h\nu \tag{14.2}$$

In order for any electrons to be emitted from a given material, the wavelength of the incident radiation must be shorter than some critical value. The frequency ν corresponding to the critical wavelength is called the *threshold frequency*.

A remarkable characteristic of photoelectric emissions is the relationship between the number and velocity of escaping electrons on one hand, and the intensity and wavelength of the incident light on the other. The maximum velocity of emission is independent of the intensity of the light, but does depend on the wavelength. An increase in intensity will result in an increase of photoelectric current, but only because a greater number of electrons are emitted. However, as long as the threshold frequency has been reached, the maximum velocity of the photoelectrons emitted from a given surface is always the same, no matter how low the intensity of the light.

Modern solar cells are made from silicon because the threshold frequency for silicon corresponds to the frequency at which solar radiation reaches a maximum value. Solar cells are made with a large junction area. The junction is made by diffusing an extremely thin p-type surface into a wafer of n-type silicon. The light falls on this face and is largely absorbed near the junction itself. The output is obtained between an ohmic contact on the p-type layer and another on the back face of the wafer. Silicon solar cells can generate an open circuit voltage up to 0.5 volts.

Well-designed solar cells can convert solar energy to electrical energy with an efficiency of approximately 10%. For most terrestrial applications the low efficiency prevents solar batteries from competing with conventional power sources. In space applications, however, multi-cell arrays of silicon batteries are used for communication satellites and space ships.

The fundamental difficulty with solar power generators lies not only with the limited efficiency of 10% for devices, but also with the amount of solar energy that can be collected. It was previously stated that the solar constant measured at the outer fringes of earth's atmosphere was 1.34×10^3 W/m^2. This means that 1 m^2 cell surface would deliver only one 1 1/3 kW power if the device were 100% efficient, 10 m^2 of collecting surface would actually be required. The condition is even worse on the earth's surface, since only a fraction of the solar constant is transmitted through the atmosphere.

As in the case of the flat plate collector and the solar furnace, diminishing energy resources coupled with increasing needs for power enhance the potential value of solar batteries. A significant amount of development activity has been undertaken to improve the efficiency of solar semiconductor devices.

14.11 CONCLUSION

Radiation phenomena are covered in this chapter. The physical mechanism for the transfer of radiated energy between two bodies is described in some detail. Concepts of blackbodies and gray bodies are introduced along with the definitions of absorptivity, reflectivity, and transmissivity.

Fundamental laws governing radiation by Kirchhoff, Planck, Stefan, and Boltzmann are illustrated. The distinction between energy emission and absorption from one surface to another and the net exchange of energy transferred from one subject to another is pointed out. Formulations for the latter type of heat transfer are developed, along with the combined effects of radiation and convection. Several types of solar systems are discussed including heat and hot water systems, solar furnaces, and solar batteries.

SOLVED PROBLEMS

PROBLEM 1: Justify the statement "Kirchhoff's law places an upper limit on the emissive power of a body."

Solution: From Eq. 14.5:

$$\alpha + \rho + \tau = 1.0$$

or

$$0 \leq \alpha \leq 1.0$$

From Kirchhoff's law:

$$\frac{W}{\alpha} = \text{constant}$$

Therefore the emissive power W is limited by the value of α. Maximum W occurs for a blackbody where $\alpha = 1.0$.

PROBLEM 2: If a blackbody is maintained at 116°C, determine (a) the wavelength at which the maximum monochromatic emissive power occurs and (b) the maximum monochromatic emissive power.

Solution:

(a) According to Wien's law (Eq. 190), the maximum $W_{b\lambda}$ occurs when

$$\lambda_{max}T = 2893 \quad \text{where} \quad \lambda_{max} = \text{wavelength in micrometers}$$

$$T = 116 + 273 = 389\ \text{K}$$

$$\lambda_{max} = \frac{2893}{389} = 7.44\ \mu\text{m or } 7.44 \times 10^{-6}\ \text{m}$$

(b) According to Planck's law [Eq. (14.12)]:

$$W_{b\lambda} = \frac{C_1\lambda^{-5}}{e^{C_2/\lambda T} - 1.0}$$

where

$W_{b\lambda}$ = maximum monochromatic emissive power

$C_1 = 3.74 \times 10^{-16}\ \text{J m}^2/\text{s}$

$C_2 = 1.44 \times 10^{-2}\ \text{m K}$

$\lambda_{max} = 7.44 \times 10^{-6}\ \text{m}$

$T = 389\ \text{K}$

$(\lambda_{max}T) = 2893 \times 10^{6}\ \text{m K}$

$$W_{b\lambda} = \frac{3.74 \times 10^{-16}\ \text{J m}^2/\text{s}}{(e^{1.44 \times 10^{-2}/2893 \times 10^{-6}} - 1.0)(7.44 \times 10^{-6}\ \text{m})^5}$$

$$= 1.14 \times 10^{8}\ \text{W/m}^3$$

Note that the monochromatic emissive power is not expressed as watts per unit volume, but rather, as watts per unit area for some particular wavelength. The results might better be expressed as

$$W_{b\lambda} = 1.14 \times 10^{8}\ \text{W/m}^2\ \mu\text{m}$$

PROBLEM 3: Determine the total emissive power for the blackbody of solved problem 2.

Solution: Total emissive power for a blackbody is given by the Stefan-Boltzmann law, Eq. (14.16):

$$W_b = \sigma T^4$$

where

$$\sigma = 5.7 \times 10^{-8} \text{ W/m}^2 \text{ K}^4$$

$$T = 389 \text{ K}$$

$$W_b = (5.7 \times 10^{-8})(389) = 1305 \text{ W/m}^2$$

PROBLEM 4: A gray body at 100°F receives radiant energy from a wall at 2000°F at a rate of 3.2×10^4 Btu/hr ft^2. Simultaneously, the body emits energy at the rate of 140 Btu/hr ft^2. What is the average absorptivity (α) of the body at (a) 100°F and (b) 2000°F?

Solution: The energy transfer rate from and to the body can be found by applying the modified Stefan-Boltzmann law [Eq. (14.18)]:

$$W = \varepsilon_g \sigma T^4$$

where ε_g represents the gray body emissivity (or absorptivity) and T is the absolute temperature of the body from which the energy is received or the temperature at which energy is emitted.

(a)

$$W = 3.2 \times 10^4 \text{ Btu/hr ft}^2$$

$$\sigma = 0.17 \times 10^{-8} \text{ Btu/hr ft}^2 \text{ °R}^4$$

$$T = 2000 + 460 = 2460\text{°R}$$

$$\therefore \quad \varepsilon_g = \alpha = \frac{3.2 \times 10^4}{(0.17 \times 10^{-8})(2460)^4} = 0.51$$

(b)

$$W = 140 \text{ Btu/hr ft}^2$$

$$T = 100 + 460 = 560\text{°R}$$

$$\varepsilon_g = \alpha = \frac{140}{(0.17 \times 10^{-8})(560)^4} = 0.84$$

PROBLEM 5: A red brick conduit 10 in square has a surface temperature of 300°F and is mounted inside a large earthen chamber whose walls are are at 50°F. Estimate the heat loss from the conduit by radiation.

Solution: This case is similar to that described in case 2 of Table 14.2. Area A_1 is the surface area of the conduit, which is completely enclosed in surface 2 and which is small compared to surface 2. The heat transfer rate due to radiation can be found from Eq. (14.23):

$$q_{1-2} = A_1 F_{1-2} F_\varepsilon \sigma (T_1{}^4 - T_2{}^4)$$

where

$$q_{1-2} = \text{h.t. rate per foot of conduit}$$

$$A_1 = (1.0 \text{ ft})(40/12 \text{ ft}) = 3.23 \text{ ft}^2$$

$$F_{1-2} = 1.0$$

$$F_\varepsilon = \varepsilon_1 = 0.93 \text{ (Table 14.1)}$$

$$\sigma = 0.17 \times 10^{-8} \text{ Btu/hr ft}^2 \text{°R}^4$$

$$T_1 = 300 + 460 = 760\text{°R}$$

$$T_2 = 50 + 460 = 510\text{°R}$$

$$q_{1-2} = (3.33)(1.0)(0.93)(0.17 \times 10^{-8})(760^4 - 510^4)$$

$$= 1400 \text{ Btu/hr per ft of conduit.}$$

PROBLEM 6: Estimate the radiation heat transfer coefficient for solved problem 5.

Solution: The radiation h.t. coefficient can be estimated by use of Eq. (14.25):

$$h_r = F_{1-2} F_\varepsilon \left[\frac{\sigma (T_1{}^4 - T_2{}^4)}{T_1 - T_2} \right]$$

$$= (1.0)(0.93) \left[\frac{(0.17 \times 10^{-8})(760^4 - 510^4)}{760 - 510} \right]$$

$$= 1.68 \text{ Btu/hr ft}^2 \text{ °R}$$

EXERCISES

1. List two significant differences between radiation heat transfer and convection heat transfer.
2. What is the wavelength of x-rays with a frequency of 2×10^{17} s^{-1}?
3. If a photon of light has a wavelength of 5000 Å, determine its energy.
4. How much energy is associated with 1 kg of matter?
5. If the total incident radiant energy on an opaque body is 2000 W/m^2, and 600 W/m^2 is absorbed, what is the reflectivity of the body?
6. If a blackbody has a temperature of 180°F, determine (a) the wavelength for maximum monochromatic emissive power and (b) the maximum monochromatic emissive power.
7. Describe the "greenhouse effect."
8. Explain how the old-fashioned blacksmith judged the temperature of a piece of metal.
9. Determine the total emissive power (W_b) for the blackbody of problem 6.
10. If the sun has a temperature of 10 000°R and a surface area of 66×10^{18} ft^2, what is the rate of radiant heat transfer if the sun is assumed to be a blackbody?
11. An oxidized copper pipe 2 in in diameter has a surface temperature of 250°F. It is enclosed in a red brick conduit 6 in square whose inside surface is at 80°F. Estimate the rate of radiant heat transfer.
12. Estimate the radiation heat transfer coefficient for problem 11.

REFERENCES

1. F. Kreith, *Principles of Heat Transfer*, 2nd ed., International Textbook, Scranton, Pa., 1967.
2. W. M. Rohsenow and J. P. Hartnett, *Handbook of Heat Transfer*, McGraw-Hill, New York, 1973.

15

REFRIGERATION AND AIR CONDITIONING

15.1 INTRODUCTION

Refrigeration can be defined as the achievement of a temperature which is lower than that dictated by the local environment. It is, in effect, a cooling process. Two broad classifications of refrigeration can be considered: natural and artificial refrigeration.

15.2 NATURAL REFRIGERATION

Natural refrigeration occurs when a substance or a region is cooled spontaneously without the use of special equipment or devices. In most practical applications, however, specially designed containers and fluid-moving devices are used to augment natural cooling processes. Natural refrigeration can be achieved in two ways:

Temperature rise of a coolant
Phase change of a coolant

Temperature Rise of a Coolant

An object can be cooled by placing it in close contact with a cooler substance; the substance is called a *coolant* and it can be a solid, a liquid, or a gas. For example, electronic components can be cooled by placing them on a cold plate, by passing cool air over them, or by immersing them in a cool liquid.

An expression for determining the quantity of heat that a coolant can absorb was discussed in Chap. 6:

$$Q = mc\ \Delta t \tag{6.19}$$

where m is the mass of the coolant, c is the mean specific heat over a particular temperature range, and Δt is the temperature rise of the coolant.

Phase Change of a Coolant

When a coolant changes phase, it absorbs a specific amount of heat from the environment so that the temperature of the environment is lowered. In Chap. 6 the quantity of heat required to effect a phase change in a unit mass of substance was designated as the heat of transformation L. The total quantity of heat removed from the environment when a mass m changes phase was shown to be

$$Q = mL \tag{6.24}$$

A classic and still widely used example of phase change refrigeration is the melting of ice. The heat of fusion for water at 1 atm pressure is about 79 cal/gm (see Table 9.6). This means that for each gram of ice that melts, 79 cal heat is removed from the environment.

One disadvantage of using melting ice for refrigeration is the need for liquid water disposal. Another disadvantage is that the cooling level is limited by the saturation temperature of water. The latter disadvantage can be diminished by use of a brine solution, that is, a mixture of salt and water. Figure 15.1 is a solubility curve for calcium chloride salt and water. Coordinates for the curve are melting temperature in Celsius degrees and weight percent composition of the constituents. The curve shows that the melting point of pure water is 0°C and the lowest melting point of the aqueous salt solution is -51°C. The lowest melting point is called the eutectic temperature, and the corresponding composition is the eutectic composition. The low melting temperatures of this brine solution make it useful in keeping highways free from ice in very cold climates. Practical

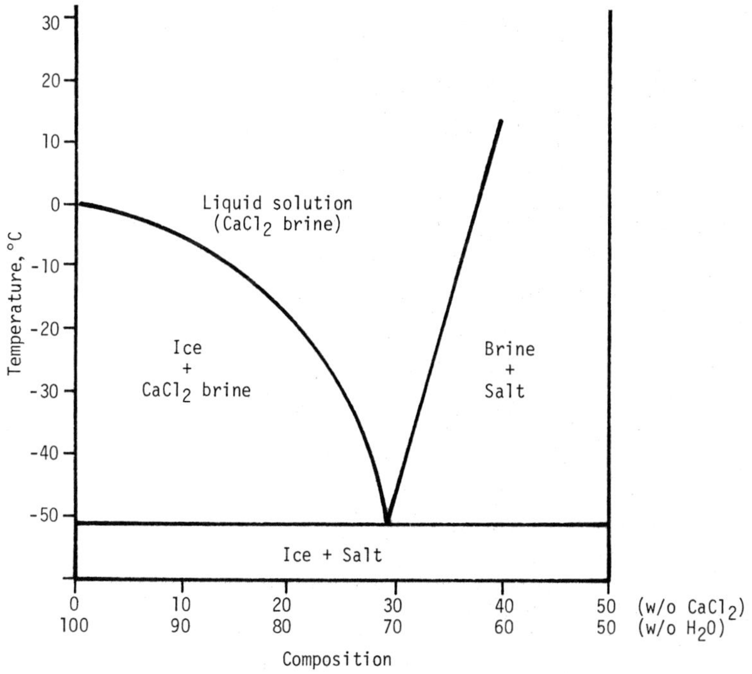

FIGURE 15.1 Solubility of calcium chloride in water. Eutectic point occurs at -51°C and 29 w/o $CaCl_2$.

cooling applications of the brine appear in refrigerated railroad cars or trucks.

Another example of refrigeration by means of a phase change is evaporation. Table 6.3 indicates that the heat of vaporization for water is 539 cal/gm. This is the reason that water stored in clay jars is kept cool. Water droplets in the pores of the jar evaporate by absorbing heat from the main body of water, so that the temperature is lowered. Human body temperature is also regulated by the evaporation of perspiration (an aqueous solution) on the surface of the skin.

Refrigeration also takes place when a substance passes directly from the solid to vapor phase. In this case, heat is absorbed from the local environment by virtue of the heat of sublimation for the coolant. A particularly useful refrigerating medium is solid carbon dioxide (dry ice). It sublimes at 79°C and 1 atm, at which point it absorbs approximately 136 cal/gm.

15.3 ARTIFICIAL REFRIGERATION

Artificial refrigeration occurs when special devices are assembled to lower the temperature of a substance or a region. Artificial refrigeration can be achieved in several ways:

Expansion of a liquid
Expansion of a gas
Electrical refrigeration
Magnetic refrigeration

Expansion of a Liquid

The temperature of a liquid is reduced when the volume of a liquid is suddenly increased (expanded). This phenomenon occurs in a *throttling process*, which is defined as a steady-flow process through a restriction in a line (Fig. 15.2). The process is essentially adiabatic because it takes place so fast that there is not enough time for any significant heat transfer; in addition, the small area of the restriction does not allow much heat transfer.

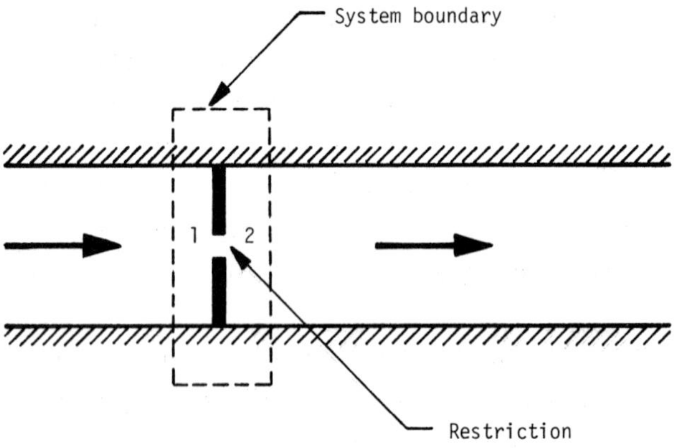

FIGURE 15.2 Throttling process.

Figure 15.2 represents a steady flow system; the first law of thermodynamics in the form of Eq. (7.10) can be applied. No work is done by a throttling process, and so $\dot{W} = 0$, and since it is adiabatic, $\dot{Q} = 0$. In addition, the change in kinetic and potential energy across the restriction can be considered negligible. Consequently, the enthalpy remains constant in a throttling process:

$$\text{For a throttling process} \quad h_1 = h_2 \tag{15.1}$$

During a throttling process a certain percentage of the liquid being throttled always flashes into vapor. This is consistent with the fact that the enthalpy of a liquid decreases as the temperature decreases. Therefore, in order for Eq. (15.1) to be valid, there must be some vaporization. This condition leads to the expansion valve equation

$$h_{f1} = h_{f2} + xh_{fg2} \tag{15.2}$$

where h_{f1} and h_{f2} are the specific enthalpies of the liquid at the inlet and outlet pressures of the expansion valve, respectively, x is the quality of the liquid-vapor mixture, and h_{fg} is the latent heat of the refrigerant at the outlet pressure of the expansion valve. Thermodynamic diagrams for a throttling process are shown in Fig. 15.3.

Expansion of a Gas

If a gas expands freely and does no work, it undergoes a throttling process. The temperature of a gas will increase, decrease, or stay constant depending on its initial temperature and pressure. A thermodynamic property called the Joule-Thomson coefficient μ determines the magnitude and sign of the temperature change. The coefficient is defined as the change in temperature with respect to pressure in a constant enthalpy process:

$$\mu_J \equiv \left(\frac{\partial t}{\partial P}\right)_h \tag{15.3}$$

From Eq. (15.3) it is evident that μ_J is the slope of a line of constant enthalpy on a temperature-pressure diagram (see Fig. 15.4).

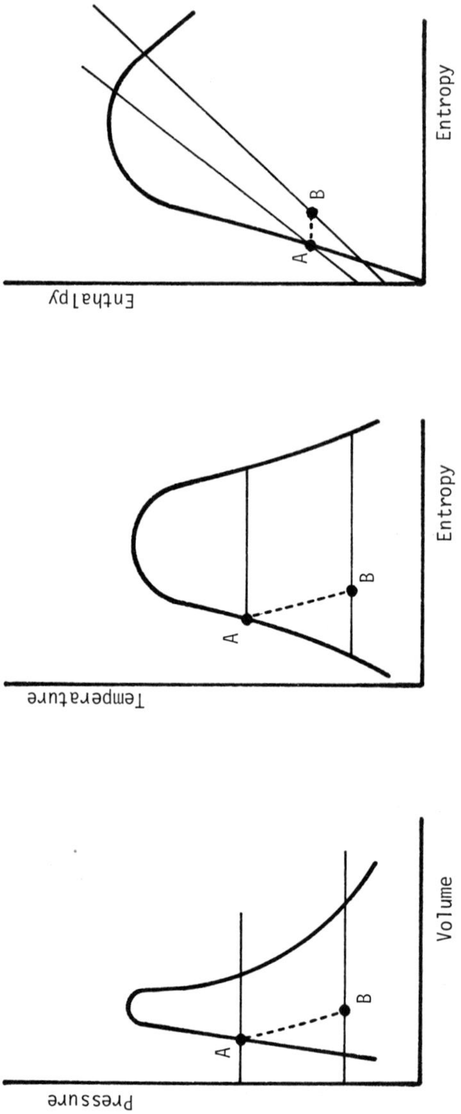

FIGURE 15.3 Thermodynamic diagrams for a throttling process.

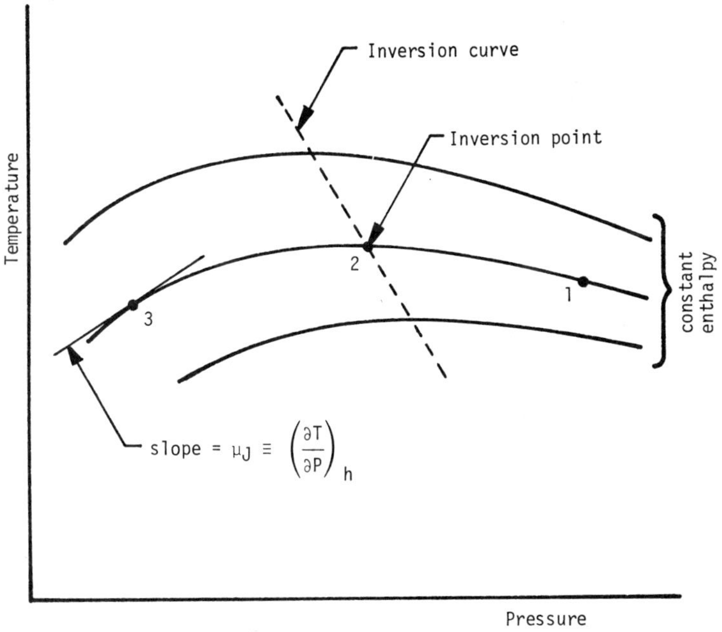

FIGURE 15.4 Slope of a temperature-pressure curve is designated as the Joule-Thomson coefficient μ_J.

Since the pressure in a throttling process decreases, the motion of the process on Fig. 15.4 is to the left. Thus, when the slope is positive, such as from point 2 to point 3, temperature decreases. When the slope is negative (points 1 to 2), temperature increases. At the inversion point where the slope is zero, there is no temperature change.

Gases such as air, nitrogen, oxygen, carbon dioxide, and ammonia have positive values of μ_J at ordinary temperatures and can be liquefied easily. Hydrogen and helium, however, have negative values of μ_J at normal temperatures and must be precooled before liquefication. The relationship can be shown by comparing the maximum inversion temperature for several gases, as shown in Table 15.1.

TABLE 15.1 Maximum Inversion Temperatures for Several Gases

Gas	Max. Temp., °C
Nitrogen	348
Air	330
Hydrogen	-78
Helium	-250

Electrical Refrigeration

An electric heat pump can be constructed from semiconductor materials, such as bismuth telluride, by utilizing the thermoelectric effects associated with the materials. The thermoelectric effects are called the Seebeck, Peltier, and Thomson effects.

Seebeck Effect: An electromotive force can be generated if heat is applied to one junction of two dissimilar conductors while the other junction remains at a fixed temperature (see Fig. 15.5). The voltage developed is proportional to the difference in temperature between the junctions, and the proportionality constant is called the Seebeck coefficient α_{ab}:

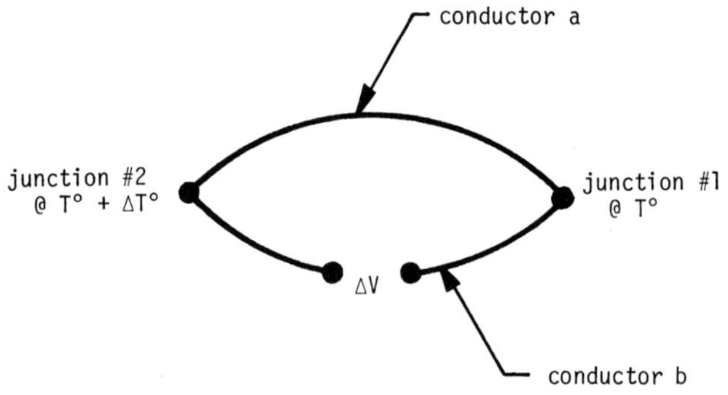

FIGURE 15.5 Schematic illustration of the Seebeck effect.

$$\alpha_{ab} \equiv \lim_{\Delta T \to 0} \frac{\Delta V}{\Delta T} = \frac{dV}{dT} \tag{15.4}$$

The Seebeck coefficient is not attributed to either material alone, but rather to the pair of materials. The change in potential may be positive or negative, depending upon the materials used. Thus, both the magnitude and sign of α_{ab} must be specified.

Peltier Effect: When a direct current is passed through two dissimilar materials, heat will be absorbed or rejected at the junction (see Fig. 15.6). It has been shown experimentally that the amount of heat transfer at the junction is proportional to the current:

$$Q_P = \pi_{ab} I \tag{15.5}$$

where the proportionality constant π_{ab} is called the Peltier coefficient. As in the case of the Seebeck effect, the Peltier effect depends on the material pairs.

Thomson Effect: If an electric current flows through a conductor on which a temperature gradient is imposed, a heat exchange will take

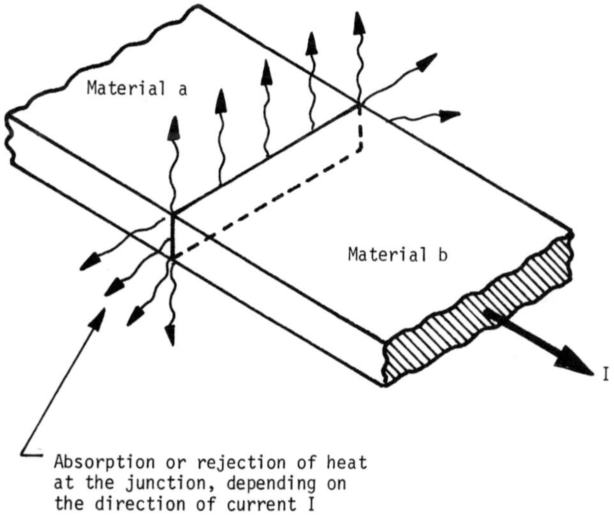

FIGURE 15.6 Schematic illustration of the Peltier effect.

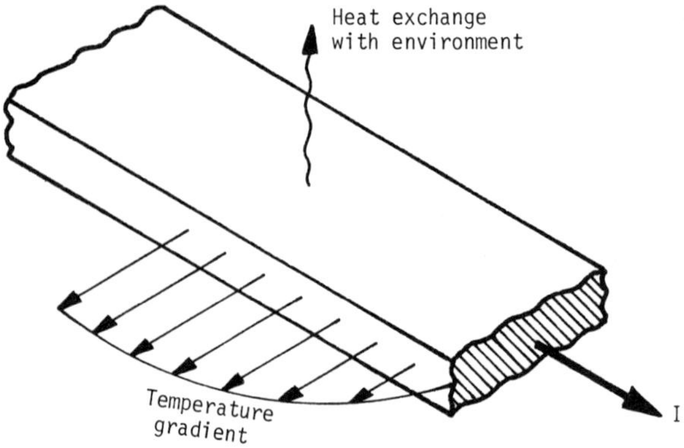

FIGURE 15.7 Schematic illustration of the Thomson effect.

place between the conductor and its surroundings (see Fig. 15.7). It has been determined experimentally that the heat absorbed or rejected per unit length is proportional to the product of the electric current and the temperature gradient:

$$Q_T = \tau I \frac{dT}{dx} \tag{15.6}$$

where the proportionality constant τ is the Thomson coefficient. Since the heat transfer in Eq. (15.10) is a second-order effect, it is often neglected in thermal analysis.

Principle of Operation: Semiconductor materials contain a distribution of free electrons. These electrons can be set in motion by the application of either a temperature gradient or an electric field. Consequently, if heat is applied to one end of a semiconductor, the kinetic energy of the free electrons will be increased, and a net electron flow toward the cold end will result. A charge is associated with each of the electrons, and so the electron flow is associated with an electric current. The dependence of the concentration of charge carriers on the temperature gradient is actually a manifestation of the Seebeck effect and is the principle on which a thermocouple operates.

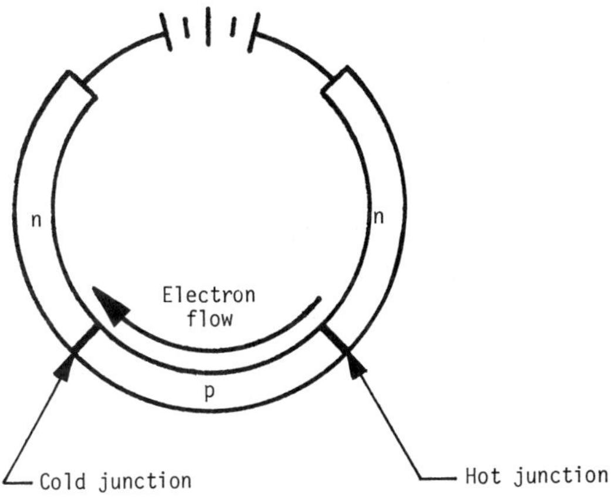

FIGURE 15.8 Hot and cold junctions between dissimilar pieces of thermoelectric materials.

Consider also three semiconductor materials connected as shown in Fig. 15.8 so that an electron flow is established through them. Since the materials are alternately negative, positive, and negative, a difference in potential energy of the charge carriers exists in each material. Consequently, in order to satisfy the first law of thermodynamics, when the carriers move across a junction, an exchange of energy with the surroundings must occur. The change in energy level establishes hot and cold junctions and makes possible the construction of a refrigerating system.

The n-type material has an excess of electrons and the p-type material has a deficiency. As an electron passes across the junction from p type to n type, it must raise its energy level, and does so at the expense of the junction. Consequently, the junction loses energy, which is evidenced by a reduction in temperature. Conversely, as the electron goes from n-type to p-type material, the junction increases in temperature. It must be remembered that electron flow is opposite to current flow.

Packaging Configurations: The configuration shown in Fig. 15.8 is obviously not convenient for a practical cooling unit. Most thermoelectric devices are constructed as shown in Fig. 15.9. Semiconductor

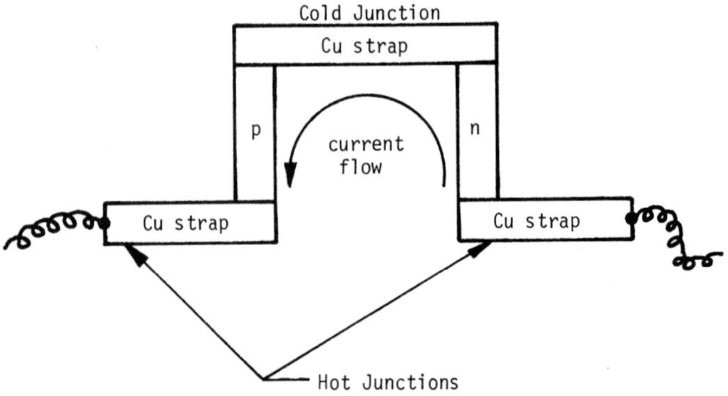

FIGURE 15.9 Typical Peltier couple.

pellets of p-type and n-type materials are soldered in series with copper straps that serve the dual function of bus bars and hot and cold junctions. The device is called a *Peltier couple*. Such a device has a relatively small cooling capacity, and so several of them must be connected together.

Comparison with a Vapor Compression Cycle: Some interesting analogies exist between vapor compression and thermoelectric refrigeration. Corresponding parts of each are shown in Fig. 15.10. The most important element in each system is the means for changing the internal energy at the hot and cold sides.

For a vapor compression cycle, the throttling valve is the device that causes the energy change. When the refrigerant leaves the condenser, it is a saturated liquid at high pressure and medium temperature. As it proceeds through the throttling valve, it is expanded adiabatically and at a constant enthalpy. Consequently, the refrigerant leaves the valve as a low pressure, low temperature, low quality vapor, and is at its lowest energy state. This makes it possible for the refrigerant to absorb the maximum amount of heat during the evaporation process. Without the throttling valve there would be a constant pressure and uniform enthalpy, and no heat pumping would occur.

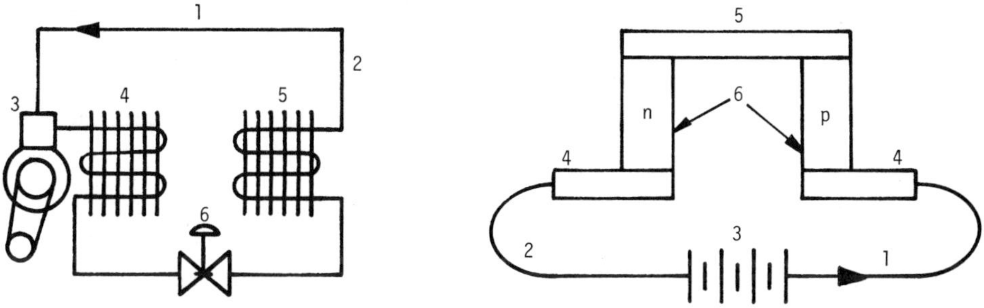

VAPOR COMPRESSION		THERMOELECTRIC
Refrigerant flow	1.	Electron flow
Leak tight tubing	2.	Electrical paths
Motor compressor	3.	Power supply
Condenser	4.	Hot junction
Evaporator	5.	Cold junction
Throttling valve	6.	Energy level

FIGURE 15.10 Refrigeration system analogies.

The analogous element in a thermoelectric refrigerating system is the difference in available electron energy in the p-type and n-type materials. No heat pumping would result if the energy levels of the electrons were the same throughout the system. Energy level diagrams for both systems are shown in Fig. 15.11.

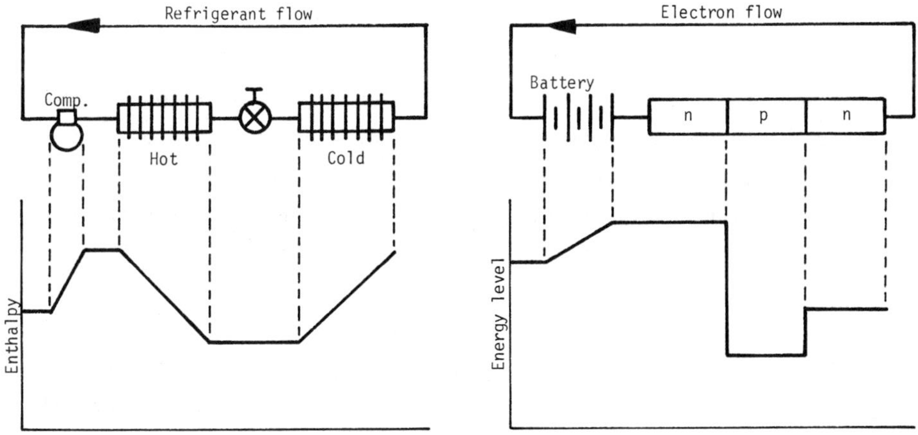

FIGURE 15.11 Energy level analogies.

Magnetic Refrigeration

Magnetic cooling is often referred to as *adiabatic demagnetization.* The process consists of initially cooling a paramagnetic salt, such as gadolinium sulfate, to a low temperature (<1.0 K) by means of liquid helium. When a magnetic field is applied to the salt, the molecules are aligned and the temperature rises. Heat is transferred to the helium until the temperature of the salt is lowered to its initial value. Finally, the salt is thermally isolated and the magnetic field is removed, resulting in a disalignment of the molecules. The disalignment process requires energy, which must come from the internal energy of the salt, and so the salt temperature drops to a very low value. Through successive actions of adiabatic demagnetization, a temperature as low as 0.001 K can be reached.

15.4 COMMERCIAL REFRIGERATION

All of the methods discussed above are used commercially for various applications. The most important, however, is the vapor compression cycle. Another refrigerating system which is useful in certain circumstances is the absorption cycle.

Vapor Compression Cycle

A vapor compression cycle is one in which a refrigerant in the form of a vapor is compressed; it is then condensed to a liquid and then throttled to a liquid-vapor mixture, which is evaporated back to a vapor. A schematic diagram of the equipment needed to complete a vapor compression cycle is shown in Fig. 8.3 in Chap. 8.

The standard vapor compression cycle is shown on temperature-entropy and pressure-enthalpy diagrams and is contrasted with the Carnot refrigeration cycle in Fig. 15.12. For the standard cycle a saturated vapor refrigerant is compressed isentropically to a superheated vapor, shown by point 2. Heat is rejected isopiestically in process 2-3 and the refrigerant leaves the condenser as a saturated liquid. An adiabatic throttling process occurs from point 3 to 4, and the refrigerant is evaporated isopiestically to the initial state.

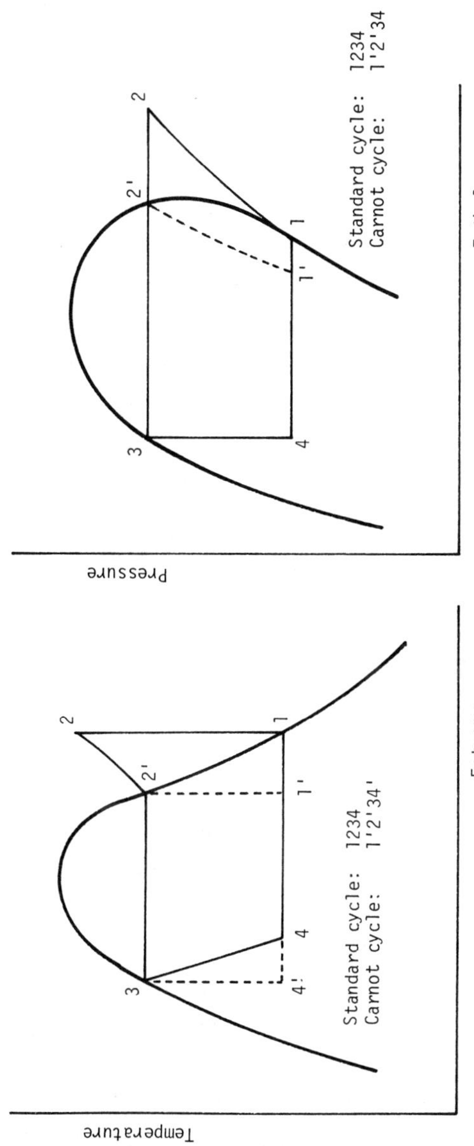

FIGURE 15.12 Standard vapor compression cycle compared with a Carnot cycle on Temperature-Entropy and Pressure-Enthalpy diagrams.

The standard cycle deviates from the Carnot cycle in the initial process because it is extremely difficult to compress a liquid-vapor mixture; also, process 3-4 is not isentropic since a throttling process is irreversible. Referring to Fig. 15.12, the processes of a standard vapor compression cycle are shown to be

Isentropic compression of a saturated vapor to the condenser pressure (1-2).

Heat rejection at constant pressure, resulting in desuperheating and condensation (2-3).

Irreversible expansion at constant enthalpy of a saturated liquid to the evaporator pressure (3-4)

Heat addition at constant pressure, resulting in evaporation to a saturated vapor (4-1).

Because of practical considerations, an actual cycle deviates from the standard cycle as shown in Fig. 15.13. It is desirable that a superheated vapor enter the compressor to ensure that no liquid is present. Also, it is best to subcool the liquid being throttled to prevent vapor bubbles from impeding flow through the expansion valve. These two conditions can be achieved by passing the liquid from the condenser and the vapor from the evaporator through a heat exchanger (see Fig. 15.14).

Another deviation occurs in the compression process, which is necessarily irreversible. Therefore, entropy will either increase or decrease during this process, depending on whether heat is transferred to or from the refrigerant. Figure 15.13 shows the increasing entropy case.

Absorption Cycle

A schematic drawing of the equipment used in an absorption cycle is shown in Fig. 15.15. The refrigerants used in such a system are ammonia or lithium bromide, which are absorbed in water. A low pressure vapor leaves the evaporator and is absorbed in a relatively weak ammonia or lithium bromide solution. Heat generated by the absorption

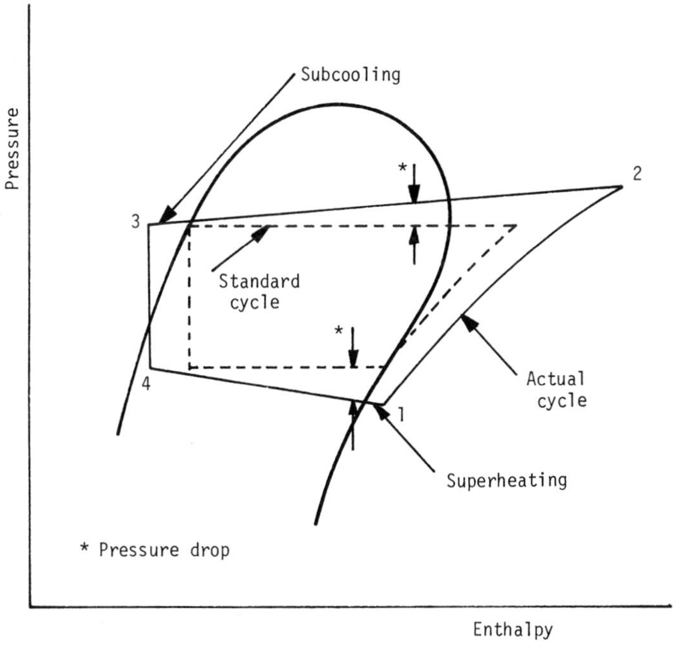

FIGURE 15.13 Actual vapor compression cycle compared with a standard cycle on a Pressure-Enthalpy diagram.

process must be transferred from the absorber. A strong solution is developed, which is pumped to the generator where a high temperature and pressure are maintained. Heat addition to the generator drives off some of the vapor from the strong solution. The vapor proceeds to the condenser, while the weakened solution flows back to the absorber. Action of the refrigerant through the condenser, expansion valve, and evaporator is the same as in a vapor compression system.

The principal difference between an absorption system and a vapor compression system is the replacement of the compressor by an absorber, pump, and generator. Replacement of the compressor means that very little input work is required by an absorption system. This advantage is somewhat offset by the need for a high rate of heat input to the generator.

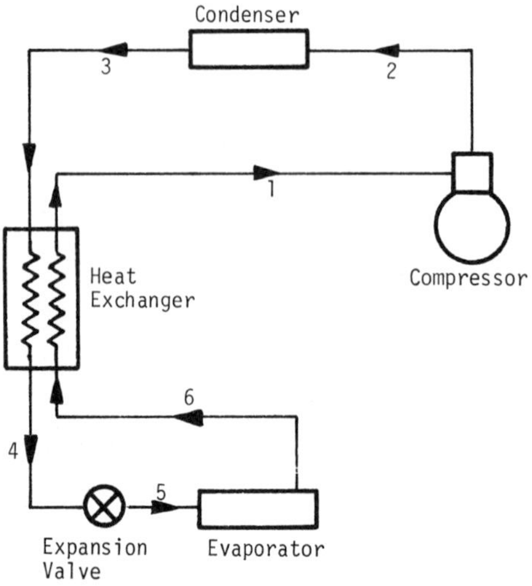

FIGURE 15.14 Schematic of actual vapor compression cycle equipment.

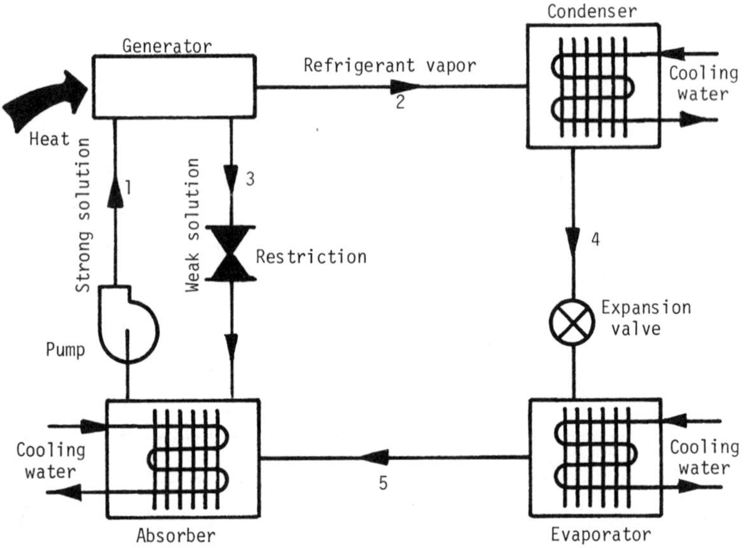

FIGURE 15.15 Schematic of a basic absorption cycle.

15.5 REFRIGERANTS

A very important element in a refrigerating system is the refrigerant. A refrigerant is the medium which absorbs heat in the evaporator and rejects heat in the condenser. A wide variety of refrigerants are commercially available to meet many different requirements. The most important commercial refrigerants are the halocarbons and the inorganic compounds.

Halocarbons are halogenated hydrocarbons, that is, hydrocarbons in which chlorine, fluorine, or bromine have been substituted for hydrogen atoms. The principal hydrocarbons used commercially are methane (CH_4) and ethane (C_2H_6). The American Society of Heating, Refrigerating, and Air-Conditioning Engineers (ASHRAE) has established a numbering system for refrigerants which has general use. Commercial manufacturers of refrigerants use the designated numbers in conjunction with proprietary names such as freon, genetron, isotron, frigen, and arctron. Rules for the numbering system are listed below:

The first digit on the right represents the number of fluorine atoms.

The second digit on the right represents one more than the number of hydrogen atoms in the compound.

The third digit from the right represents one less than the number of carbon atoms in the compound. This digit is omitted if it is zero.

Table 15.2 is a partial listing of halocarbons. A fuller listing can be found in publications of ASHRAE.

Refrigerants used in the early days of refrigeration were inorganic in nature, and some are still widely used. ASHRAE has designated three figure numbers, beginning with a 7, to be used for inorganic refrigerants, as shown in Table 15.3.

15.6 PSYCHROMETRY

Atmospheric air is never completely dry, but always contains a certain amount of water vapor. *Psychrometry* is the study of how water vapor

TABLE 15.2 Halocarbon Refrigerants

No.	Chemical Name	Formula
10	Carbontetrachloride	$C\ Cl_4$
11	Trichloromonofluoromethane	$C\ Cl_3\ F$
12	Dichlorodifluoromethane	$C\ Cl_2\ F_2$
22	Monochlorodifluoromethane	$CH\ Cl\ F_2$
50	Methane	CH_4
113	Trichlorotrifluoroethane	$C_2\ Cl_3\ F_3$
125	Pentafluoroethane	$C_2\ H\ F_5$
170	Ethane	$C_2\ H_6$
218	Octafluoropropane	$C_3\ F_8$
290	Propane	$C_3\ H_8$

is present in air at any particular time. A knowledge of psychrometry is important for anyone interested in refrigeration or air conditioning systems.

For most psychrometric analyses, atmospheric air can be considered to be a mixture of two ideal gases. Technologists are generally interested in several different kinds of temperatures, pressures, densities, and humidities. The latter term, humidity, represents a measure of how much water vapor is in the mixture. Problems can be

TABLE 15.3 Inorganic Refrigerants

No.	Chemical Name	Formula
717	Ammonia	NH_3
718	Water	H_2O
727	Air	--
744	Carbon Dioxide	CO_2
744A	Nitrous Oxide	N_2O
764	Sulfur Dioxide	SO_2

solved by a direct thermodynamic analysis, or in a much simpler manner, by use of a psychrometric chart.

Thermodynamic Analysis

The temperature and pressure of water vapor in atmospheric air can be shown conveniently on a temperature-entropy diagram such as Fig. 15.16. The water vapor is considered to be superheated and is shown as state 1, with partial pressure P_w. Temperature T_{DB}, *the dry bulb temperature*, is the air temperature as measured with an ordinary thermometer. State 2 represents the saturated vapor state at the dry bulb temperature and fixes P_s, the saturated pressure of the water vapor.

The saturated temperature corresponding to the partial pressure P_w is called the *dew point temperature*, T_{DP}, and is shown as state 3 on the diagram. An intermediate temperature known as the *wet bulb temperature*, T_{WB}, is indicated at state 4. T_{WB} is measured with an

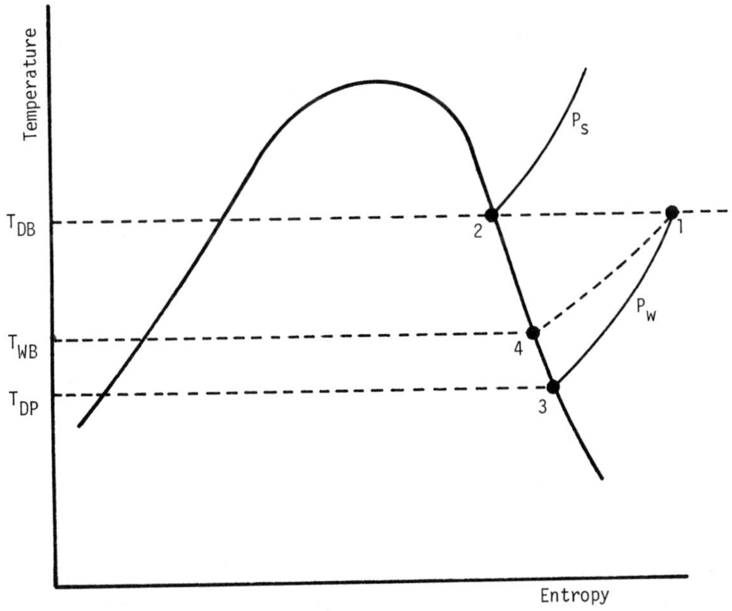

FIGURE 15.16 T-s diagram for water vapor in atmospheric air.

instrument called a psychrometer, which contains a thermometer with a water saturated cloth wick around the bulb. When air is moved past the wick, water evaporates and the temperature reading is lowered. When equilibrium is established, the temperature attained is read as T_{WB}. Psychrometers also include standard thermometers for a simultaneous reading of T_{DB}. Obviously, the wet bulb temperature will depend on how much water vapor exists in the air stream.

Atmospheric pressure P is made up of the dry air partial pressure P_a, and the water vapor pressure P_w. According to Dalton's law for partial pressures [Eq. (9.23)]:

$$P = P_a + P_w$$

Relative humidity ϕ is defined as the ratio of the partial pressure of water vapor in the air mixture to the saturation pressure of the water vapor at the same temperature:

$$\phi \equiv \frac{P_w}{P_s} \tag{15.7}$$

Specific or absolute humidity ω is defined as the ratio of the mass of water vapor to the mass of air:

$$\omega \equiv \frac{m_w}{m_a} \tag{15.8}$$

but since the ideal gas law applies to both dry air and water vapor, it can be shown that

$$\omega = \frac{P_w}{P_a} \tag{15.9}$$

Psychrometric Chart

Psychrometric calculations can be greatly simplified by making use of psychrometric charts. Numerous versions of these charts are available but, essentially, they all give the same information. A widely used version is the chart developed by ASHRAE, of which a skeleton diagram is shown in Fig. 15.17.

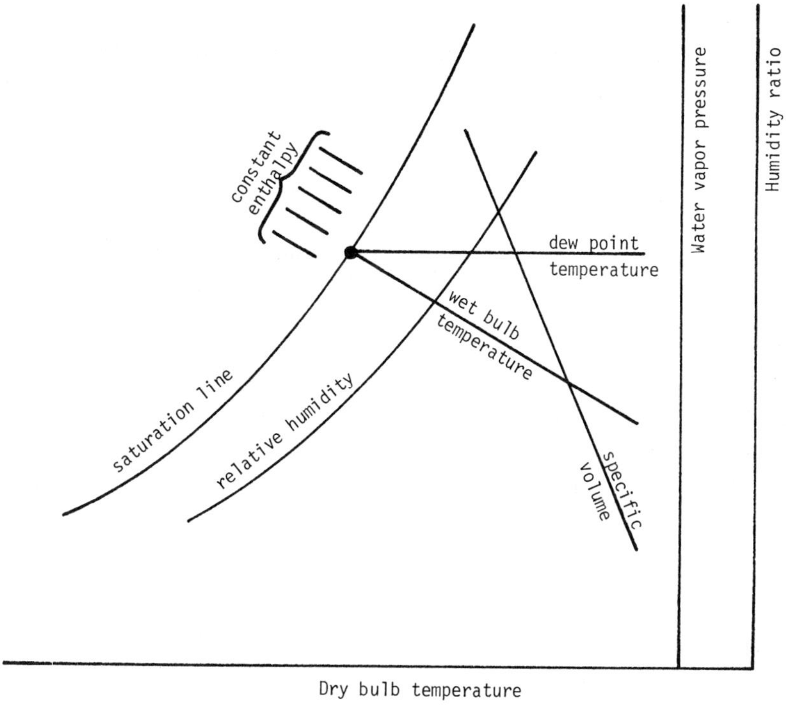

FIGURE 15.17 Skeleton diagram of the ASHRAE psychrometric chart.

The standard psychrometric chart has an abcissa designating dry bulb temperature and an ordinate designating absolute humidity. Wet bulb temperature lines, sloping upward to the left, are shown at 1° intervals, while horizontal lines, at the same interval, represent dew point temperatures. The saturation curve, on which wet bulb and dew point temperatures are shown, curves upward to the right. Note that on this curve the wet bulb, dew point, and dry bulb temperatures are all equal.

Relative humidity lines at 10% intervals have a shape similar to the saturation curve. Lines showing constant enthalpy are almost parallel to the wet bulb temperature lines. The specific volume of dry air is shown by straight lines sloping upward to the left. The ASHRAE chart also includes a monograph which shows the ratio of changes of sensible and total heat, and enthalpy and absolute humidity.

Standard charts are drawn for normal atmospheric pressure (29.92 inHg or 760 mmHg). These charts can be used as long as the pressure does not vary by more than ±4%. Beyond these values, analytical techniques must be used or revised charts must be available.

15.7 AIR CONDITIONING

Conditioning of air generally implies a control of its temperature, humidity, and velocity, as well as control of dust, bacteria, and odors. A schematic representation of an air conditioning system is shown in Fig. 15.18. Air enters the system with absolute humidity ω_1 and leaves with ω_2. The system can be humidified or dehumidified by adding or removing water in the amount $\omega_2 - \omega_1$.

The system can also be heated or cooled by the addition or removal of heat at a rate $\dot{Q}$. A typical air conditioning system is one of steady flow, and so the first law of thermodynamics in the form of Eq. (7.10) can be applied. Since no work is done and changes in potential and kinetic energies can be neglected, the general equation for an air conditioning system is

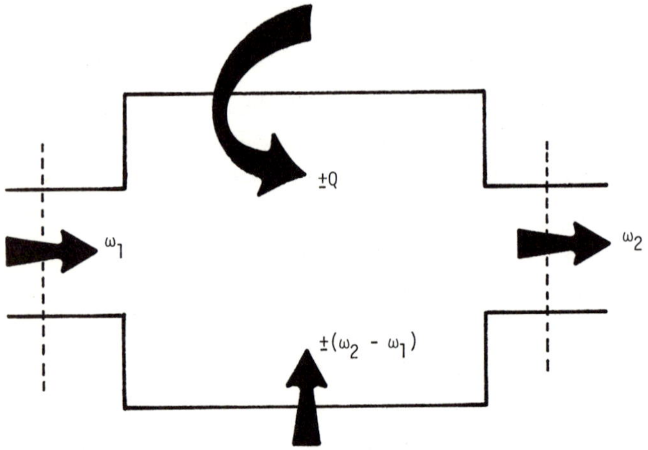

FIGURE 15.18 Schematic diagram of an air-conditioning system.

$$\dot{m}h_1 \pm \dot{Q} = \dot{m}h_2 \tag{15.10}$$

The refrigerating effect Q, that is, the heat transferred per pound mass of refrigerant, can be expressed as

$$Q = h_2 - h_1 \tag{15.11}$$

The capacity of an air-conditioning system is expressed in *tons of refrigeration.* The term reflects the early days when cooling capacity was given in terms of tons of ice melting per unit time:

1 ton of refrigeration ≡ 12 000 Btu/hr

The performance of a refrigerating or air-conditioning system can be expressed as the coefficient of performance $(COP)_R$, which was introduced in Chap. 8.

$$(COP)_R = \frac{Q_L}{W} = \frac{Q_L}{Q_H - Q_L} \tag{8.2}$$

15.8 CONCLUSION

In this chapter several methods of refrigeration are described. For example, natural refrigeration is shown to occur by means of a temperature rise or a phase change in a coolant. Artificial refrigeration is shown to occur by means of an expansion of a liquid or vapor, or by electrical means. In the latter case, the Seebeck, Peltier, and Thomson effects are explained. A comparison is made between vapor compression and thermoelectric refrigeration.

A number of different refrigerants are discussed, along with the standard numbering system used to identify them. The determination of the amount and state of water vapor in air is shown to be covered by the discipline of psychrometry, and an explanation is given on how to read a psychrometric chart.

SOLVED PROBLEMS

PROBLEM 1: Estimate how much the temperature is lowered when steam at 7 bar and 260°C is throttled to 1.4 bar. What is the approximate Joule-Thomson coefficient?

Solution:

(a) From metric steam tables (superheated steam): h at 7 bar and 260°C = 2974.9 J/gm and for a throttling process

$$h_1 = h_2$$

also

$$T = 251°C \text{ for superheated steam at 1.4 bar}$$

Therefore, temperature is lowered 260 - 251 = 9°C

(b)

$$\mu_J \simeq \frac{\Delta t}{\Delta p} = \frac{9°C}{7 - 1.4} = 1.61 \text{ °C/bar}$$

PROBLEM 2: A vapor compression refrigeration system, using Refrigerant 12, works between an evaporator temperature of 10°F and a condenser temperature of 110°F. The pumping rate for the refrigerant is 180 lbm/hr. Determine (a) the refrigerating effect, (b) the coefficient of performance, and (c) the capacity of refrigeration in tons.

Solution: Use refrigeration tables and a pressure-enthalpy diagram for Refrigerant 12 [5]. Also, refer to Fig. 15.11 for general T-s and p-h diagrams for a refrigerating system.

State 1 is saturated vapor at 10°F:

$$t_1 = 10°F$$

$$p_1 = 29.335 \text{ psia}$$

$$h_1 = 78.335 \text{ Btu/lbm}$$

$$s_1 = 0.16798 \text{ Btu/lbm°R}$$

State 2 is superheated vapor at the saturation pressure for the condenser temperature; process 1-2 is isentropic:

$$p_2 = 151.11 \text{ psia}$$

$$s_2 = s_1 = 0.16798 \text{ Btu/lbm°R}$$

$$h_2 = 91 \text{ Btu/hr}$$

State 3 is saturated liquid at 110°F:

$$h_3 = 33.531 \text{ Btu/lbm}$$

Since process 3-4 is a throttling process:

$$h_3 = h_4 = 33.531 \text{ Btu/lbm}$$

(a)

$$\text{Refrigerating effect} = Q_L = h_1 - h_4 = 78.335 - 33.531$$
$$= 44.8 \text{ Btu/lbm.}$$

(b)

$$\text{COP} = \frac{Q_L}{-W}$$

where

$$-W = \text{work supplied to compressor}$$
$$= h_2 - h_1 = 91 - 78.335 = 12.7 \text{ Btu/lbm}$$

$$\text{COP} = \frac{44.8}{12.7} = 3.5$$

(c)

$$\text{Capacity} = \frac{(44.8 \text{ Btu/lbm})(180 \text{ lbm/hr})}{(12\ 000 \text{ Btu/hr}) \text{ per ton}} = 0.67 \text{ ton}$$

PROBLEM 3: A room has a dry bulb temperature of 70°F, a relative humidity of 60%, and a pressure of 1 atm. Using a psychrometric chart [5], determine (a) wet bulb temperature, (b) dew point temperature, (c) enthalpy, and (d) specific humidity.

Solution: Locate the state described by drawing a vertical line from the abcissa at 70°F and noting the intersection with the curved line representing 60% relative humidity.

(a) Moving upward to the left along a line of constant wet bulb temperature, read the value 61°F.

(b) Moving to the left along a horizontal line, read the dew point temperature at 56°F.

(c) Moving upward to the left along a line of constant enthalpy (parallel to the constant wet bulb temperature), read the enthalpy as 27.1 Btu/lbm.

(d) Moving horizontally to the right, read the specific humidity as 63 grains of moisture, or 0.0095 lbm H_2O per lbm of dry air.

EXERCISES

1. Identify two ways in which cooling processes are developed in nature.
2. Estimate the freezing point of a mixture of water and calcium chloride in which there is 20 w/o $CaCl_2$.
3. Why is a throttling process adiabatic in nature?
4. Estimate the temperature reduction when steam at 100 psia and 400°F is throttled to 20 psia. What is the approximate Joule-Thomson coefficient?
5. Explain the basic operating principle for thermoelectric refrigeration.
6. Explain the basic operating principle for vapor compression refrigeration.
7. Explain the basic operating principle for magnetic refrigeration.
8. What is the major advantage of an absorption refrigeration cycle over a vapor compression cycle?
9. An ammonia vapor compression refrigeration system operates between an evaporator temperature of 20°F and a condenser temperature of 60°F. Determine (a) the refrigerating effect, (b) the COP, and (c) the capacity in tons of refrigeration.
10. Air at 1 atm pressure has a dry bulb temperature of 80°F and a wet bulb temperature of 65°F. Determine (a) the relative humidity, (b) the enthalpy, (c) the dew point temperature, and (d) the specific humidity. (Use a psychrometric chart.)

BIBLIOGRAPHY

Carrier Air Conditioning Co., *Handbook of Air Conditioning System Design*, McGraw-Hill, New York, 1965.

Handbook of Fundamentals, ASHRAE, New York, 1967.

Seely, J. H. and R. C. Chu, *Heat Transfer in Microelectronic Equipment*, chap. 8, Marcel Dekker, New York, 1972.

Stoecker, W. F., *Refrigeration and Air Conditioning*, McGraw-Hill, New York, 1958.

Strock, C. and W. B. Foxhall, *Handbook of Air Conditioning, Heating and Ventilation*, Industrial Press, New York, 1959.

Systems and Equipment Guide and Data Book, ASHRAE, 1967.

INDEX